The Art of the Bee

The Art of the Bee

Shaping the Environment from Landscapes to Societies

ROBERT E. PAGE, JR.

OXFORD
UNIVERSITY PRESS

OXFORD

UNIVERSITY PRESS

Oxford University Press is a department of the University of Oxford. It furthers
the University's objective of excellence in research, scholarship, and education
by publishing worldwide. Oxford is a registered trade mark of Oxford University
Press in the UK and certain other countries.

Published in the United States of America by Oxford University Press
198 Madison Avenue, New York, NY 10016, United States of America.

Library of Congress Cataloging-in-Publication Data
Names: Page, Robert E., author.
Title: The art of the bee : shaping the environment from landscapes to societies /
Robert E. Page Jr.
Description: New York, NY : Oxford University Press, 2020. |
Includes bibliographical references and index.
Identifiers: LCCN 2019058018 (print) | LCCN 2019058019 (ebook) |
ISBN 9780197504147 (hardback) | ISBN 9780197504161 (epub)
Subjects: LCSH: Bees—Behavior. | Social behavior in animals.
Classification: LCC QL569.4 .P34 2020 (print) | LCC QL569.4 (ebook) |
DDC 595.79/915—dc23
LC record available at https://lccn.loc.gov/2019058018
LC ebook record available at https://lccn.loc.gov/2019058019

1 3 5 7 9 8 6 4 2

Printed by Sheridan Books, Inc., United States of America

Contents

Preface

Alexander von Humboldt explored South America at the turn of the 19th century, 1799–1804. For 5 years he crisscrossed New Spain from Venezuela to Peru, exploring the river systems of the Amazon, scaling the highest peaks of the Andes; and with his traveling companion, Aimé Bonpland, he recorded the natural history with extensive collections of plants, 4,000 pages of diary notes, and paintings of the wonders that he encountered. A product of the Age of Reason and the Enlightenment, he was a polymath. His knowledge of natural history and all dimensions of science at the turn of the 19th century was staggering, as were his fundamental contributions, founding several new scientific disciplines that are still identifiable today. He published his observations in a massive five-volume collection of lectures, *The Cosmos*. The *Personal Narrative* of his travels, published in 1814, marked him as the greatest explorer-scientist of his time and served as the inspiration for Charles Darwin on his nearly 5-year circumnavigation of the earth aboard the HMS *Beagle*.

Views of Nature, a condensed version of parts of *Cosmos*, was his most acclaimed book and is a most remarkable work. Chapters are built around themes, presented as "views" with the double meaning of seeing something and to have an opinion. Within each chapter, and indeed within each paragraph and even within individual sentences, Humboldt weaves observations of geology, geography, botany, zoology, ecology, atmospheric science, and even anthropology in a seamless way. Humboldt appealed to artists to learn about nature and ecology and paint it. He believed that artists and writers could do more to advance an understanding of science and nature than the scientific specialist. His plea was for making science understandable to the public, a plea for popular science. Edwin Church took up his challenge and painted his wonderful *Heart of the Andes* after retracing Humboldt's steps in South America. When you look at the 1.7 by 3 meter canvas, first exhibited in 1859 and now housed in the New York Metropolitan Museum of Art, you are drawn in by the size and the detail of the painting, reflecting the complex ecology of a mountain stream in the heart of the Andes, one that does not exist. Church did not reproduce an actual location but built an ecologically accurate composite of a region of the Andes. Within the painting he weaves a picture of the geology, botany, ecology, and anthropology of the region. The painting draws you closer as you view it and find ever more exquisitely intricate detail.

Twenty-five years ago, my friend and mentor Harry Laidlaw wanted to write a honey bee biology textbook. He wanted me to coauthor the text, so we sat down and mapped it, making outlines for each chapter. The outline formed the typical kind of textbook, stratified layers of knowledge assembled by level of biological organization: biogeography, systematics, anatomy, physiology, behavior, then followed by apiculture, to be excavated from front to back. When the outline was finished, it looked very much like the excellent book by Mark Winston *The Biology of the Honey Bee*, published in 1987 by Harvard University Press. I decided we didn't need another one, and we still don't. But I do think we need a different kind of honey bee biology book. Like Humboldt, I wanted to write a scientific book that isn't hierarchical in its organization but instead weaves basic scientific information around interesting themes, paints a picture like Church that provides a view defined by the chapter titles.

I want the reading experience to be more like learning biology from individual case studies than the normal textbook that consists of strata that are excavated one layer at a time. My target reader isn't the specialist, though I think he or she would find this book interesting, but the person who has a basic knowledge of biology and a fascination with bees, perhaps an educated hobby beekeeper (there are a lot of them) or an undergraduate or graduate student with an interest.

My good friend Bert Hölldobler was instrumental in my being awarded a Carl Friedrich von Siemens Foundation Fellowship for 2017–2018. Bert encouraged me to write a book about the honey bee as a superorganism during my fellowship year. I had deliberately avoided using the term *superorganism* in most of my own work for most of my career; I worried about what it really meant as a metaphor or model for social evolution, and it had ended up encompassing almost everything to some sociobiologists and philosophers of biology and almost nothing to others. In the past few years, Bert has convinced me that it is a useful concept that has truly framed and defined much of what we do in insect sociobiology and convinced me that my own research is an example of how superorganisms evolve. I decided not to write a book about the superorganism, but it occupies half of this one.

The chapters were originally written as stand-alone essays, each containing basic biology built around my views of individual themes. The chapters weren't intended to weave seamlessly together. The presentation of biology is the prime objective; the themes are designed to keep the reader interested. The concept of views in the sense of Alexander von Humboldt is the unifying theme. I split three of the original essays into two parts each, to aid the reader in navigating the information (Chapters 1–2, 5–6, 7–8).

As I wrote it, the book logically self-organized into natural history (Chapters 1–3 and 9) and social evolution (Chapters 4–8, Epilogue). It begins with a treatment of bees as landscape artists. I take on one of the challenges of Humboldt, to look at natural history from the eyes of an artist. I am not an artist, but in Chapters 1 and 2, I attempt to paint a mental view of the earth being transformed by the bees as they became locked into a coevolutionary dance with flowering plants some 120 million years ago, taking the earth from being dominated by brown and green to a kaleidoscopic array of color and how they're still painting our landscapes today. In Chapter 3, I present a view of honey bees as environmental engineers, engineering and building their own protective nest environment and, through their pollination activities, engineering the niches of many other plants and animals with which they share the environment. Chapter 3 is the longest because it contains the most biology. I tried to divide it into two chapters but failed.

Chapters 4–8 are views of social insect colonies as superorganisms. Chapter 4 views insect societies through a lens of political philosophy. I make a loose comparison of the social organization observed in insects with social contracts of human societies. I discuss the evolution of altruism in the context of a social contract written by the social history of populations onto the genetic makeup of individuals. Chapters 5 and 6 follow with an exploration of the superorganism metaphor, how it has been defined and used, where it fails, and my view of its current usefulness. Chapter 7 dissects the honey bee superorganism into the mechanisms that are responsible for transforming an aggregation of individual bees into a colony that functions like an organism. In Chapter 8, I present the results of my own research where I used human-assisted selection as an analogue of natural selection to get a glimpse at how superorganisms can evolve.

Chapter 9 provides a description of the most important process of a superorganism, reproduction. In this chapter, a honey bee colony swarms and produces a new replacement queen that must mate and become the repository of the germ line of the colony. It ends with the birth of a new worker, the link between two superorganism generations.

Humboldt had extensive notes at the end of each chapter to substantiate his claims. I don't. Instead, I've listed many papers and books I used as resources. All scientific claims I present, including statistics and specific data that are presented, are fully supportable and were checked and verified. When it's my view, my opinion, I own it. Most of what I present is central dogma of the discipline, though that's not always the case, in which case I point it out. And finally, I realize I fall far short of Humboldt; but then, that's no surprise.

I'm indebted to many. First, I thank Bert Hölldobler. Bert has been a great friend, colleague, and mentor and especially a strong supporter of my endeavors and career. I'm indebted to the Alexander von Humboldt Foundation for the Humboldt Research Prize I received in 1995 as well as the Carl Friedrich von Siemens Foundation for the fellowship I received in 2017. They first introduced me to Humboldt and opened many doors for me over the years, doors that brought me opportunities to work in Germany and in the United States with many outstanding colleagues. The Siemens Foundation gave me a year free from all worries and responsibilities so that I could focus on writing this book. I also thank the sponsors and organizers of the annual Falling Walls Circle in Berlin, Germany. In 2017, those of us who attended were challenged to think about how we could help bridge the information gap between science and public understanding. I pledged to try to do that in this book. I hope I succeeded.

Richard Platt read an earlier manuscript and offered suggestions and much encouragement. Richard is a retired forester and hobby beekeeper, part of my target group. Peggy Coulombe helped me with the editing and prepared the manuscript for submission.

1

Environmental Artists

I perhaps owe having become a painter to the flowers.

Claude Monet (date unknown)

We're moved by the colors and images of the 19th-century impressionist painter Claude Monet. His vibrant landscapes filled with wildflowers and re-creations of his estate gardens draw us in, revealing his brushstrokes that resemble blobs and smears of vital pigments extracted from the paints on his palette. Like Monet, bees paint our environment—the landscape their canvas, flowering plants their pigments, their foraging behavior the brushstrokes. I was raised in the Central Valley of California. Every spring the foothills of the Sierra Nevada are painted gold, purple, and yellow with poppies, lupines, mustard, and asters, resembling an impressionist canvas, a consequence of 250 million years of coevolution of plants and their insect pollinators.

1.1 Coevolution of Flowering Plants and Bees

Prior to the rise of flowering plants, the landscape was dull. The first plants invaded land more than 450 million years ago. Species of ferns, horsetails, clubmosses, and other "primitive," non-seed-producing, non-flowering plants dominated. The earth was painted from a palette of green and brown. The first seed plants were gymnosperms, which originated around 100 million years later and eventually gave rise to the conifers that dominated the earth with massive forests. The first flowering plants, the angiosperms, appeared perhaps as early as 250 million years ago. The rise of flowering plants resulted in a burst of new plant species as they adapted to their insect pollinators, mostly beetles and short-tongued flies. Think of how the landscape changed. From non-seed-producing plants with flowerless, unidentifiable organs of reproduction and wind-pollinated conifers with reproductive cones to a planet covered in brightly colored flowers. But the change didn't take place at once. At first, the flowers were tiny, not brightly colored, and contained in structures like the catkins found today in willows, birches, and oak trees. From the beginning, the new flower-making plants depended on insects to pollinate them. Insects had been around for more than 200 million years longer than flowering plants and were plentiful and ready to eat the floral and vegetative parts and, at a significant cost to the plant, pollinate them.

However, the burst of new flowering species was small compared to the one that occurred 125 million years ago. It was marked in the fossil record with a new kind of pollen, one with more pores through which the pollen grain germinates and grows a tube through which it shares genetic information with the ova, producing a fertile seed. It also occurred around the time of the first bees, a dependency that produced a burst of new species of both bees and plants. Within a short time, the flowering plants replaced the conifers as the dominant trees on earth; and today, there are estimated to be more than 350,000 species, of which 74%–94% require some kind of animal pollination. More than 16,000 species of bees have been described of an estimated 25,000 believed to exist. Of those, the vast majority live solitary lives, only about 6% are social, the most important of which, to humans, is the honey bee.

The burst of new species of flowering plants and bees that occurred 125 million years ago created great difficulties for Charles Darwin. In an 1879 letter to his botanist friend Joseph Dalton Hooker, he called it an "abominable mystery." Darwin's theory of evolution by natural selection depended on (a) variation in traits, such as floral structures and pollen grain morphology; (b) variation that in some way must have an effect on the survival and/or reproductive success of the individuals with those traits, enabling them to leave more offspring; and (c) heritability—in other words, the parents must transmit the traits to their offspring. If these three conditions are met, then the most favorable traits will increase in their representation in the population from one generation to the next—the population will evolve. For Darwin, this was a process that took place in tiny steps, at glacial speeds, over geological time. The presence of the bursts of new species that occurred "suddenly" in the fossil record gave some credit to the arguments of the creationists of his time and cast doubt on his gradual evolution arguments. Darwin resolved the difficulty, at least to his own satisfaction, by invoking coadaptation. The burst of new species was a consequence of the evolution of the intimate relationships between plants and their insect pollinators. He believed that this could greatly accelerate the speed with which new species evolved.

Floral Adaptations

The vast array of species and floral structures that we see today represents adaptations to different animal vectors of pollen and the various ways plants have adapted to the same vectors. They use color, shape, and odor to advertise their presence and recruit pollinators; and they use nectar and pollen as carbohydrate and protein rewards. They also optimally position their stamens with their pollen-laden anthers and the pistil, with its sticky stigma prepared to capture pollen grains, optimally to effect fertilization of their ova and to spread their genes via pollen to others. For example, the bilateral symmetry seen for many flowers, such as garden peas and snapdragons, is likely to have evolved in response to bee visitations (Figure 1.1). Their petals provide a platform for bees to land and probe their nectaries and, through this activity, transfer pollen onto the bee and from the bee to the stigma of the flower—pollination.

Nature's palette, from which colors are drawn to paint today's earth landscape, spans the color spectrum from ultraviolet to red, including violet to yellow. This is the visual spectrum of bees. Their color vision differs from ours in that they can see ultraviolet (we can't) and we can see red (bees can't). Bees have three kinds of color-sensitive photoreceptors. Each has a different light frequency where it's maximally sensitive. Experiments have shown that with these photoreceptors honey bees can easily discriminate among the colors ultraviolet, violet, blue-green, yellow, and purple, a combination of yellow and ultraviolet. Our own purple is likewise drawn from the extremes

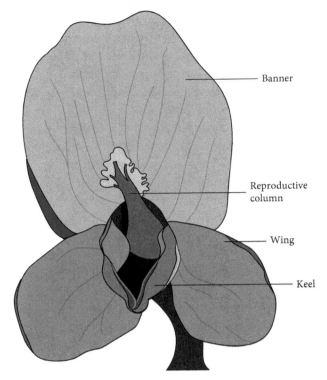

Figure 1.1 Diagram of alfalfa blossom. Alfalfa flowers use bees as their primary pollinators. The trip mechanism is an adaptation that showers the bee with pollen and at the same time picks up pollen that's on the body of the bee to effect its own pollination. The specialized parts for pollination are the reproductive column and the keel. The reproductive column with the pollen-containing anthers, male parts, and the style that contains the female reproductive parts are held captive and spring-loaded in the keel until released by a bee when it probes the nectaries. The other petals, the banner and wings, have been modified to attract visiting bees.

of our spectrum of color vision by mixing red and blue. The color vision of bees preceded the angiosperms and so did not arise by the insects becoming adapted to the floral colors following the angiosperm species explosions. It's ubiquitous throughout the Hymenoptera—ants, wasps, and bees—and appears to be of ancient origin. Plants adapted to the color vision of bees and even use ultraviolet reflectance, invisible to us, to guide foragers to the sugar-rich nectaries.

Let's take a moment to ponder our world as we see it. Color does not exist. The world is colorless. Color is a product of our own creation, painted within our minds by electrical impulses from our eyes. Light is electromagnetic (EM) energy that strikes our earth, primarily coming from our sun; it's a continuum of waves of different lengths, only some of which fall within our ability to see them, visible light, roughly 380–800 nanometers. Below our visible range are X-rays and ultraviolet, and above are infrared, microwaves, and radio waves, all part of the EM spectrum. We see discrete colors from violet to red because we have different photoreceptors in the retina of our eye that are tuned to absorb EM energy across specific frequency ranges and convert it into electrical impulses that get integrated into what we perceive as color. Objects in our environment absorb EM energy of some wavelengths and reflect back the rest, some of which then enter our eyes and, if they're the right length, paint images within the occipital lobe, the visual cortex of our brain. Beauty is truly in the eye, and brain, of the beholder.

Researchers have mapped the reflectance of flowers across natural landscapes on multiple continents and have repeatedly shown that the color distribution of flowers matches the color sensitivity of bees. They do not match with birds, butterflies, or flies, other important plant pollinators. This clearly demonstrates that bees are the Monets of the natural world. They're primarily responsible for the evolution of the floral shapes and colors, and they pollinate the plants today, making them responsible for the seeds that are sown on the landscape canvases around us.

Adaptations of Bees

Bees also became adapted to flowers. The honey bee is an exemplar of adaptation of anatomy, physiology, and behavior. The coadaptation of flowers and honey bees took place over millions of years in a dialectical dance of seeking a solution that maximizes the pollination of the plant and the resource exploitation of the bees.

Anatomical
Bees show special anatomical adaptations for collecting nectar and pollen. The proboscis of a bee is derived from modified mouthparts, the paired maxillae and the labium. They form a straw-like apparatus for probing nectaries and extracting nectar. Bees are covered by branched, plumose hairs. This is a

characteristic of all species of bees and aids in the collection of pollen. As bees fly through the air, the hairs can become electrostatically charged and, when on a flower, attract loose pollen from the anthers, like a magnet. Pollen also gets trapped in or adheres to the branched hairs as the bee probes for nectar or actively engages in pollen collection. Once the bee is covered in pollen, she uses her legs (insects have three pairs of legs) to brush the pollen from her body and transfers the pollen to the specially modified rear legs. Each rear leg has a rake, a pollen press, and a corbicula (Figures 1.2 and 1.3) on the outside of the leg, where the pollen is packed into the dense balls observed on pollen foragers.

Physiological

The principal sugars found in nectar are sucrose, glucose, and fructose. Honey contains primarily glucose and fructose. It's necessary to convert sucrose into those two sugars as part of the process of making honey. Honey bees have a special enzyme, invertase, to accomplish this. They also produce enzymes that they add to honey that result in the production of acids that help preserve the honey. Pollen is used as a source of protein, but its hard outer shell is difficult to digest, impossible for humans. Pollen grains have special pores that can be digested, releasing the protein contained inside; but it requires a special enzyme. Bees produce this enzyme.

Learning

Plants evolved flowers with special shapes, colors, and odors that are attractive to bees. Honey bees evolved the ability to learn the shape, color, odor, and location of flowers and the time of day that they yield their sugar and protein rewards. The combination of traits links bees and flowering plants in a special relationship. Plants require pollen from flowers of the same species to be deposited on the stigma. They need pollinators that show fidelity to single floral species, at least on a single foraging sortie. Honey bees learn the characteristics of a flowering plant and show high foraging fidelity for specific species or varieties, fulfilling the requirements of the plant. However, this breaks down with obligatory cross-variety fertilization, as is found in commercial crops like almonds. Almond trees come in different varieties. Seed (nut) production requires that the pollen and the flower making the seed be from different varieties, thus insuring cross-pollination. When you pass a commercial almond orchard in bloom, it's easy to see that there's more than one variety represented in the orchard. You can tell by the color of the

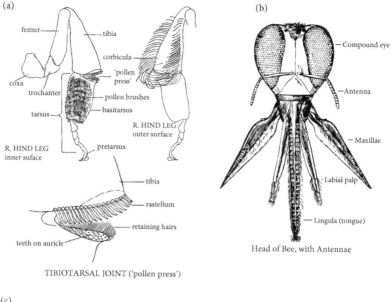

Figure 1.2 Adaptations of bees to flowers. Panel A: Adaptations of the hind leg for brushing the pollen from the body and packing it into the pollen baskets for transport from the flowers to the nest. From *Anatomy and Dissection of the Honey Bee* (plate 7), by H. A. Dade, 1977, International Bee Association. Reproduced by permission of the International Bee Research Association (ibra.org.uk). Panel B: Adaptation of the mouthparts into the proboscis of the bee—part tongue, part straw—used in probing the nectaries of flowers and sucking out the nectar. From *The Honey-Bee: Its Nature, Homes and Products* (fig. 27, p. 98), by W. H. Harris, 1884, Religious Tract Society (https://www. biodiversitylibrary.org/page/21494767#page/119/mode/1up). In the public domain. Panel C: Plumose, branched hairs that cover the body of a bee. Branched hairs are adaptations for collecting pollen from the flowers. Pollen grains get entangled in them or cling to them as a consequence of differences in static electrical charges, then are brushed off and packed into the pollen baskets. From "Physiology and Social Physiology of the Honey Bee," by E. Southwick, in J. Graham (Ed.), *The Hive and the Honey Bee* (fig. 1, p. 172), 1992, Dadant and Sons. Reprinted with permission from Dadant and Sons.

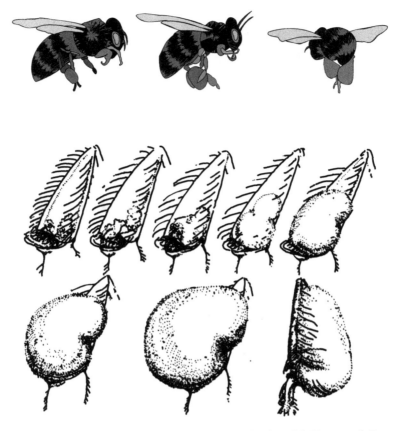

Figure 1.3 Honey bee loading its pollen baskets (corbicula). Top panel: Bees load their pollen baskets while hovering over the flower. They "wallow" in the pollen on the anthers of the flower, then lift off and hover. The forelegs are used to brush the pollen off of the body. Pollen is transferred to the middle legs, where the bee uses the brush to move it to the pollen baskets on the hind legs. The bee then presses the hind legs together and packs the pollen into the baskets. Bottom panel: The progressive packing of the corbicula. From "Activities and Behavior of Honey Bees," by N. E. Gary, in J. Graham (Ed.), *The Hive and the Honey Bee* (pp. 323 and 324), 1992, Dadant and Sons. Reprinted with permission from Dadant and Sons.

flowers, often alternating with rows of trees. Bees can tell the difference too and tend to stick to one variety as they forage, thus reducing pollination because they deposit the pollen of previous trees they've visited onto non-receptive stigmas of flowers of the same variety. It's believed that when bees return to the hive, they rub against each other and transfer pollen between foragers that visited different flowers. When they return to the orchard, they transfer pollen between varieties, effecting pollination.

Honey bees learn by doing, a process called *operant* or *tacit learning*. As they approach a flower, they perceive location, color, shape, and odor and know the time of day. They alight on the flower and taste the sugar in the nectaries as they probe for nectar. Bees have taste receptors on their tarsi (feet) as well as their proboscis (tongue). As a consequence, they associate the features of the flower with the sugar reward. This is known as *associative learning*. The key element to this kind of learning is the *forward pairing* of the conditioned stimulus, in this case the characteristics of the flower, and the nectar reward. The bees see and smell the flowers before they taste the nectar. *Reverse pairing*, where the bee gets the reward first and then gets the stimulus, doesn't work. Bees don't make the appropriate association. We do. It matters little to us if we see the color of the Oreo cookie before or after we taste it; we quickly associate the taste with the round shape and black color.

Many laboratory studies have shown that the bee's perception of the sugar reward affects learning. Like us, they place a higher value on what they perceive as more rewarding and learn it faster, like remembering one's debit card pin number. Sugar perception can be measured by a proboscis extension response assay. Bees are placed in small tubes to restrict their movements, with their heads extending out of the tube. A small droplet of sugar solution is then touched to the antenna. If the concentration of the sugar is high enough, the bee will reflexively respond by extending the proboscis (Figure 1.4). It's a reflex, like when the doctor hits your knee with a mallet and your leg extends; you can't not do it. By offering bees a series of sugar solutions of increasing concentration you can determine their response threshold to sugar. That is the lowest concentration to which they respond. Bees with lower response thresholds perform better on laboratory tests of associative learning. They learn faster. They put more value on a given concentration offered as the reward than do bees with higher response thresholds. Perception of the sugar reward varies widely among individual bees as a consequence of genetic variation, age, and experience.

Navigation

As "central place foragers," bees fly out from the nest site and explore the surrounding environment in search of food resources. They return to the nest with the resources they collect. To do this, they need to be able to navigate

Figure 1.4 Proboscis extension reflex of a worker honey bee. The bee is restrained in a tube. A droplet of sugar solution is touched to the antennae, releasing a reflexive extension of the proboscis (tongue). Reprinted from "The Development and Evolution of Division of Labor and Foraging Specialization in a Social Insect (*Apis mellifera* L.)," by R. E. Page et al., 2006, *Current Topics in Developmental Biology, 74*, 253–286, fig. 2 by Joachim Erber. © 2006 with permission from Elsevier: https://www.sciencedirect.com/science/article/pii/S007021530674008X.

out and find their way back. To aid them, they have a toolkit of navigation mechanisms. First, they have a compass that depends on the location of the sun. As light from the sun passes through the atmosphere, it becomes polarized. The pattern of polarized light in the sky depends on the angle of the sun relative to where you are looking. Bees have special sensors in their eyes for detecting the polarized light patterns. On cloudy days, they can't see the sky; but they can still locate the sun using ultraviolet light detectors. Ultraviolet light penetrates cloud cover, allowing bees to use the location of the sun as a navigational marker. With heavy clouds, bees can get to and from a resource by relying solely on landmarks that they learn; otherwise, they stay home until the weather changes.

However, as the earth turns, the sun is always changing location relative to the horizon, making it an unreliable marker unless you know the time of day, and bees do. They learn the movement of the sun across the sky and reference it to an internal clock. We know they have the clock because we can train them to forage at specific times of day. If you anesthetize a bee, you can stop her clock. When she awakens and takes a foraging trip to a learned foraging station, her flight path will be offset by the amount of time lost. In other words, she will misinterpret the direction based on the current location of the sun by the amount of time she was anesthetized.

Another tool in the navigational toolkit is an odometer. Bees can determine how far they have flown, a skill important for returning to the nest and revisiting a profitable resource. The odometer is derived from what is called the *optomotor response* of their visual system. Each compound eye of the bee is composed of about 6,900 individual, hexagonally shaped facets called *ommatidia*. Each facet has its own lens and pigments that focus light onto a light receptor (retinula) cell. As a result, the bee eye is extremely sensitive to movement. As objects pass by the eye, they successively trigger responses from the individual facets. As a bee travels to and from a resource, she measures the optic flow past her eyes that results as she passes objects in the environment. High optic flow indicates a greater distance, low flow, a shorter distance. Bees can be tricked into thinking they flew greater or shorter distances by manipulating the patterns of objects in their flight paths.

The odometer plus the ability to determine a flight vector (direction and distance) from a given landmark along a resource flight path, using their sun compass and internal clock, give bees the basic tools for navigation. The last tool in the toolkit is a path integrator that combines the compass and odometer information. There are two views on how their navigation system works from these basic tools. One view is that a bee learns and remembers a specific set of landmarks and vectors that lead to a resource and back to the nest. She travels from landmark to landmark, updating the path vector at each one, and arrives at her destination. I use a similar system when navigating in a new environment. For example, I leave my hotel, turn left, go to the first intersection, walk past the convenience store, and the restaurant is the next building. I remember the landmarks and run them in reverse to get back to the hotel. I don't yet have a mental map of the neighborhood. The paths to each resource are independent. The alternative view is that the bee has an integrated map, a cognitive map, of the environment that she has explored. The map results from combining all of the learned individual paths to resources, resulting in a global representation of the foraging environment, like the experience-based mental maps that we use to find our way. We're able, for example, to imagine in our minds all of the alternative routes we can take to go to the grocery store in our neighborhood even though we have never traveled them. We overlay familiar routes onto a larger mental map. Experimental requirements to determine if bees have a cognitive map, like we have, are very stringent and continue to be challenging for investigators. If we could only become a bee for a day, we would know.

Dance Language

The most amazing adaptation of honey bees is their dance language, through which they convey to nest mates the direction and distance to where they're foraging. In 1973, Karl von Frisch shared the Nobel Prize for Medicine or Physiology for deciphering that complex language. He showed that it's an abstract, symbolic language that's second only to human language in the animal world. That still holds, at least as far as we have been able to decipher the languages of any other animals, perhaps a problem with our deciphering abilities rather than their absence of language. Bees accomplish this with tiny brains with about 900,000 neurons compared with our 100 billion. The story of how von Frisch unlocked the dance language and the orientation abilities of bees over a 60-year period is fascinating and serves to explain how it works.

In 1923, von Frisch published his first paper on the dance language of bees. It was known as his *olfactory theory*. Von Frisch noted, as many before him going back at least as far as Aristotle, that he could put out a dish with honey and it might take several hours before a bee discovered it. But soon thereafter, many bees would be mobbing the food source; they were recruited. To study this phenomenon, he used an observation beehive so that he could observe the activities of bees inside the nest. An observation hive has one layer of frames stacked on top of another, which are sandwiched between glass plates. In a normal hive, multiple combs are organized parallel to each other, and it's impossible to observe the activities of bees on interior combs without greatly disturbing their activities. He set his observation hive in a field under a covered shelter. In his early studies, he trained bees to feeding stations 10 meters (m) from the hive. This way he could observe bees both at his feeder and in the hive. He noticed early on that returning foragers performed two kinds of "dances" on the comb. The foragers returning from his feeding stations at 10 m performed what he called a "round dance," while others performed what he called a "tail-wagging dance," also known as a "waggle dance." He assumed that the round dance was performed by nectar foragers because all of the bees visiting his sugar solution feeders performed this dance, and the waggle dance was performed by pollen foragers. He noted that many of the bees doing the waggle dance had pollen loads on their hind legs. Pollen foragers will often perform the dance when they return to the nest, then go unload their pollen into cells and return to the dance floor and continue to dance. He thought that those bees performing the waggle dance without pollen loads had already unloaded their pollen.

Initially, von Frisch provided sugar solution with a small amount of honey. The honey provided an odor to the solution. He arranged feeding stations around the observation hive at the four points of the compass. Recruited foragers showed up at all stations. His conclusion was that the round dance stimulated bees to fly out and look in all directions. However, he added odor to his test, at first by putting flowers next to some stations but not others and later by using essential oils placed on cardboard under the feeders. The external body of a bee is called the *exoskeleton* and is covered by a waxy layer to which odors cling. Bees visiting flowers, or feeding stations provided with odors pick up the odors and transport them back to the nest. Von Frisch found that when he provided odors at some stations but not others, the recruited bees only showed up at those stations with odors matching the training station. From these studies, he published his olfactory theory of recruitment in 1923. His conclusion was that the dances performed by returning foragers stimulated bees to go out and search for floral resources with the same odors as those clinging to the body of the dancing forager or contained in the nectar they collected.

Von Frisch suspected that bees were able to convey information about the distance to the resource. In 1944, he set up a new experiment where he trained bees to two feeders: one at 10 m, the other at 300 m. He found that when he provided food at 10 m, foragers performed round dances and that the vast majority of recruits showed up at 10 m (174:10 at 10 m and 300 m, respectively). When he provided food at 300 m, the foragers performed waggle dances, and the vast majority of recruits went there (8:61). He concluded that he had been wrong; the two dances weren't about nectar and pollen. Round dances were performed when the food was near the hive and did not contain specific distance information, while waggle dances were performed for greater distances with distance information. He subsequently showed that the round dance transitions into a waggle dance between 20 m and 100 m, going through various intermediate forms (Figure 1.5). His experimental bees were a geographical subspecies commonly known as carniolans. The majority of commercial bees used in North America are subspecies known

<20m   >100m

Figure 1.5 Waggle dance of the honey bee. Top: Diagram of a waggle dance with workers in attendance. The dancer moves on the vertical comb in a figure-eight pattern as indicated by the arrows. Each half-cycle of the figure eight, the bee moves her abdomen back and forth in a waggle movement while she moves in a straight line through the figure eight. She alternates turning left or right after each straight run. Bees attending the dance follow the dancing bee on the vertical comb, in the dark nest, and obtain information about the distance and direction to the flowers. Bottom: Transitions in the recruitment dance. Bees foraging close to the nest perform a round dance that contains no specific distance or directional information. As the foraging distance increases, the dance transitions to a full waggle dance. The transition distances are different for different geographical "races" of bees, a kind of geographical language dialect.

as Italian bees. Italian bees make the transition from round dance to waggle dance over a much shorter distance; hence, different subspecies "speak" different dialects.

The distance information in the waggle dance is contained in the duration of the straight run portion of the dance. This is the portion where the bee shakes her abdomen back and forth. There are many other correlations with the distance, such as sound bursts detected by bees through surface vibrations (bees lack organs for hearing) and the total time to make a complete figure-eight cycle; but the duration captures them all and is easy for us to measure with a stopwatch. Von Frisch was able to decode the distance by increasing the feeding station distances incrementally and measuring the changes in the dance. Greater distances have longer durations (0.5 second = 300 m; 1 second = 500 m; 1.33 seconds = 1,000 m; 2 seconds = 2,000 m).

The discovery that the waggle dance also contains directional information came next. Von Frisch put out feeders and scent-marked cards at 200 m around his observation hive. When he supplied food at one location, the trained, marked foragers performed a waggle dance; and the recruited bees showed up at the corresponding feeder, not the other locations. When he changed the direction of the feeder, the bees trained to that feeder did a waggle dance, and the recruits then showed up at it and not the others. He noticed that the straight-line portion of the waggle dance, that part where the bees waggle the abdomen and go up the middle of the figure eight, changed its angle on the comb for the different feeding stations. Keep in mind that these dances are being conducted on a vertical surface of the comb. Von Frisch deduced that the changing angle of the dance must in some way indicate different directions. He then fed trained bees at a feeding station located 200 m north of the observation hive. He kept the one feeding station supplied all day and watched the dances of the returning trained foragers. They performed the waggle dance, as expected; but the angle of the dance changed throughout the day. He deduced that the angle of the dance was changing at the same rate as the change of angle of the sun above (Figure 1.6). Then, he had it. He had deciphered the dance language of the honey bee.

As the sun traverses the sky, the angle from the food source to the sun changes. Imagine you are standing on top of the hive (Figure 1.6). The angle between the sun and the food source at any moment is found by first dropping a line from the sun perpendicular to the horizon. Point one arm toward that point on the horizon and the other toward the food source. The angle made by your two arms lies on a horizontal plane and defines the angle of the food source from the sun. But how does this get translated to the vertical combs on which the bees dance, and how does it compensate for the constantly changing movement of the sun? In their dance, the bees use the top of the comb, straight up, to represent the current position of the sun. This does not change as the sun moves; straight up always represents the location of the sun. The angle of the straight-line portion of the dance from the vertical indicates the direction to the food source. For example, if bees are foraging in the opposite direction of the sun, the straight-line part of the dance will point down on the comb. If the source is 90 degrees to the right of the sun, then the

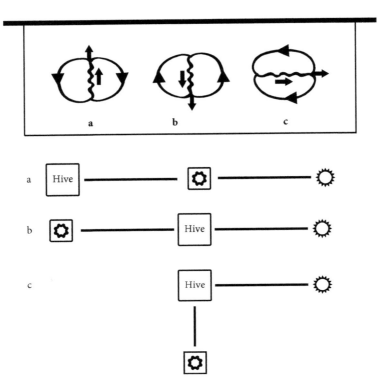

Figure 1.6 Interpreting the waggle dance. Top: Explanation of the symbols used. Middle: Diagrams of dances on the comb corresponding to the foraging directions shown below. The dances are performed on the vertical surface of the comb, shown here as a frame of comb removed from a hive. The arrows indicate the direction of the dancing bee, and the zigzag line indicates the waggle performed for the straight-line portion of the dance. For example, in (a) the bee is indicating straight up on the comb, (b) straight down, (c) 90 degrees right. Bottom: Diagrams of direction of food resource, the foraging station, relative to the position of the hive and the sun. The dancing bee and the bees that attend the dance interpret the dance relative to the position of the top of the vertical comb. The top of the comb represents the current direction of the sun that's outside the nest and not visible. The direction of the sun is relative to the horizon. Illustration a: The foraging station is located in the same direction as the sun relative to the hive. The waggle dance indicates straight up on the comb. Illustration b: The foraging station is in a direction opposite of the sun relative to the location of the hive. The forager dances straight down on the comb. Illustration c: The foraging station is located 90 degrees to the right of the sun relative to the hive. The forager's dance indicates 90 degrees to the right of the top of the comb.

dance will point to the right of the comb, etc. Bees use their solar compass and internal clock, parts of their navigational toolkit, to compensate for the movement of the sun while they perform or attend dances.

1.2 Honey Bees in the New World

Honey bees aren't native to the New World. The first records we have of honey bees in the New World have them coming by boat from England and landing near Jamestown, Virginia, in 1622. There is some evidence that they were brought earlier by Spanish priests, though no direct records exist. The honey bees escaped their domiciles, became feral, and soon began spreading across the North American continent. By 1900, they were coast to coast. Since they aren't native, they can't have played a formative role in the evolution of our native plant species. They aren't the engineers of the native plant habitats of North America.

The native plants of North America coevolved with solitary bees and bumble bees, another group of social bees. Some relationships of native solitary bees and specific native plants are extremely interdependent. The bees require a specific kind of plant, and the corresponding plants have evolved characteristics that favor pollination by the specific bees. Such is the case of the so-called squash bees and some plants in the squash family. But even though honey bees weren't present during the coevolution of plants and bees in North America, they still make a significant contribution to the pollination of native plants and plants used in agriculture. They're still painting landscapes of North America.

Pollination of Agricultural Crops

Honey bees pollinate about one third of crops and are believed to contribute up to $55 billion to the economy of the United States. This is a difficult number to derive considering the broad reach of agricultural commodities that include the supply chain to the table. Seed alfalfa is a good example of the far-reaching effects of honey bees, including seed production, growing and baling the alfalfa, the effect of alfalfa on the beef industry (or leather shoes!), and the realization that cows fed alfalfa, pollinated by bees, produce milk that gets made into ice cream that is sold at a baseball game.

Almonds

Driving up the Central Valley of California in March, more than 640 kilometers (km) of almond orchards line the highway from Bakersfield to Red Bluff. Kilometer after kilometer of trees and the ground below them are covered in white and pink blossoms. Honey bees have transformed this valley over the past 30 years by being effective pollinators of an economically important agricultural crop. Today, there are more than 400,000 hectares (ha) of almonds under cultivation in California, more than double the number 30 years ago, with 280 trees per hectare, a total of about 112 million trees. Each tree yields about 7,000 almonds after producing about 28,000 flowers. At 7.5 million flowers per hectare, there are about 3 trillion flowers produced annually that yield about 750 billion almonds. Almonds are entirely dependent on bees for pollination. The flowers are self-incompatible and must be pollinated with pollen derived from a different variety. Bees are the essential vectors of pollen needed for cross-pollination. The recommended number of colonies per hectare for effective pollination is five. As a consequence, 2 million hives, about 60 billion bees, are moved into almond orchards during the winter to provide pollination services. In 2016, there were about 376,000 ha in production, and the average yield was 2,591 kilograms (kg) per hectare, with earnings of about $5.37/kg to the grower, for a total cash crop of about $5.2 billion. The impact of bees on the agricultural environment is enormous.

Alfalfa

You can see the pollinating activities of bees on different timescales. In the preceding section, I showed how bees have changed the landscape of the planet on a geological timescale and the agricultural landscape in decades. With their recruitment system and ability to navigate over 300 km^2, they can change the landscape on a timescale of hours.

A honey bee lands on the keel of an alfalfa blossom. The keel, looking like a landing platform, is really a trigger for the spring-loaded stamens and style-laden pistil (Figure 1.1). Two petals compose the keel. The weight of the bee and her probing action to locate the sweet nectary separate the petals of the keel, thus tripping the flower by releasing the reproductive organs that fly up, hitting the bee in the head, and the anthers explode, propelling pollen over the back, the stigma prepared to accept pollen from another plant from the back of the forager. Alfalfa flowers require cross-pollination. The nectaries are now open for business to be exploited by this and subsequent bees that visit but only temporarily. The flower, now tripped, will rapidly fade from purple-blue to pink and wilt, nothing more to offer and no longer advertising for a pollinator.

You can see the pollinator activities across a field of seed alfalfa in the morning. Beehives are typically placed in clusters within the field. New flowers that matured overnight are worked by the bees as soon as it's light enough and warm enough in the morning. Over the course of the morning, the alfalfa field near the colonies will turn from purple to pink as the tripped flowers senesce. Bees will work the flowers closest to the cluster of hives first. That can be observed by looking over the field: close to the hives it looks pink, and it becomes darker and richer in color at greater distances until you reach the zones of overlap of foragers between clusters of hives. This can also be seen by seed yields. Plants within 100 m of the hives produce 30% more seed than those 300 m away.

Honey bees aren't efficient pollinators of alfalfa. The most efficient trippers of flowers are the pollen foragers, usually fewer than 30% of all foragers. Foragers have an aversion to being slapped by the reproductive parts of the flowers. They will instead learn to avoid alfalfa and go to other competing floral sources if they're available even though a field of alfalfa can secrete more than 2,000 kg of nectar/ha. Nectar foragers learn to avoid tripping the flowers by probing the nectaries from the side, often using holes in the flowers that were cut by other solitary bees that might be foraging in the field. However, honey bees have a major impact on seed yield when enough colonies are made available, increasing yields manyfold. Seed yields can be as high as 1,000 kg/ha. Studies done in large horticultural cages have shown an increase from 13 kg/ha without bees to 900 kg/ha when honey bees are present.

It's obvious that the honey bee has changed the landscape of agriculture. She has partnered with the farmers and growers in determining what crops are profitable and will be planted. The reach of the honey bee is extended, worth billions of dollars annually, reaching from seed production to your dinner table. But agriculture has likewise had a huge effect on the honey bee, mostly negative.

The Decline of Honey Bees

"A thousand little cuts" is how Mark Winston describes the worldwide decline in honey bee populations. The alarm was rung in 2006 when beekeepers reported losing more than one third of all their hives. The pattern was the same: the colony looked healthy on last inspection, and then just a few weeks later, it was dead or there were only a handful of bees and a queen, no trace of the other thousands of bees. The syndrome was called *colony collapse disorder* (CCD), and it threatened the very survival of the commercial bee industry. Commercial beekeepers are losing 850,000 of the 2.6 million colonies in the United States, annually. The number of commercial beekeepers shrank from about 210,000 in 2002 to about 120,000 in 2013.

Commercial beekeeping is large scale, with most beekeepers having thousands of hives, some more than 50,000. As a consequence, management includes standard inputs of food supplements (corn syrup for nectar and pollen substitutes or supplements for protein), miticides to control for predatory mites, antibiotics to control brood diseases, and antiprotozoal drugs for dysentery. One hundred twenty-one different pesticides, fungicides, and herbicides have been found in combs from US commercial honey bee colonies—each, by itself, below toxic levels and, combined, not reaching thresholds for major concern. However, recent research has shown that the effects of these residual chemicals and the various diseases interact, making them much more dangerous in combination than predicted from the individual effects. The tendency of scientists working on this problem, or beekeepers suffering from CCD, is to try to lay the blame on one new factor (reminds me of the "one thing" as proclaimed by Curley, the rugged cowboy in the 1991 film *City Slickers* [Castle Rock Entertainment]), such as a new agricultural pesticide or a newly introduced disease agent. However, it's becoming clearer that, indeed, the loss of commercial colonies is death by a thousand little cuts provided by current beekeeping and agricultural practices.

Bees are loaded on trucks, transported thousands of miles following pollination contracts or nectar flows, and set down on the ground, where they're exposed to the cocktail of agricultural chemical practices and monoculture food resources. Agricultural environments lack the diversity of pollen with its diverse proteins and amino acids that are important for a healthy colony. On top of all of this direct challenge to bee health, beekeepers present an epidemiologist's nightmare by maintaining bees on pallets, four colonies to a pallet, and setting up dense populations of colonies. Diseases, predators, and parasites spread quickly from colony to colony due to their close proximity and through the deliberate exchange of bees and combs between them.

A general belief is that hosts, like honey bee colonies, become resistant to their pathogens and parasites and that the pathogens and parasites become less virulent—they reach a truce with the host, where they weaken them but don't kill them. This is a coevolutionary process between host and disease agent. However, pathogens and parasites normally have much shorter life cycles than their hosts; they reproduce much more often. It's more likely that the pathogens and parasites evolve strategies to manipulate their hosts and maximize their transmission from host to host.

Commercial beekeeping practices have selected for diseases, predators, and pathogens to be more virulent and harmful to colonies. Temperate honey bees have a generation time of about 1 year. The mite *Varroa destructor*, formerly called *V. jacobsoni*, is the most important pest of honey bees—delivers the biggest and deepest cut of CCD—and has a generation time of less than 20 days. The mites can evolve much faster than bees, placing bees at a disadvantage in the host–parasite evolutionary arms race. *Varroa* mites spread through a population by weakening colonies so that infected bees are unable to defend themselves from the honey-robbing activities of other colonies. (Yes, honey bees are marauding thieves when given a chance.) The mites jump on the robber bees and hitch rides back to their colony, where they begin to reproduce and start the process over. *Varroa* mites should weaken a colony but not kill it before they can be transported to another; otherwise, they die with the colony and become locally extinct. Honey bee colonies in "the wild" are spread out, not grouped together in tight bundles of sometimes hundreds of hives as you find with commercial bees. Discovery of a weakened colony by a stronger one may take time. The best strategy is to weaken them slowly and wait. However, commercial colonies are packed close together, and it's easy to discover those weakened by mites and rob them. The best strategy for mites in commercial apiaries is to weaken them quickly and transport them to other adjacent hives. Dead colonies are replaced by beekeepers, so the supply of new hosts is unlimited. Mites should evolve to be highly virulent in commercial apiaries, and they have. Studies of the buildup of mite populations in commercial hives in California have shown population growth rates 30 times higher than in temperate climates without intensive commercial beekeeping.

Surveys of feral honey bee colonies in California demonstrated the devastating effects of *Varroa* on colonies. It was discovered in California in 1989. A 1990 survey of 208 wild, cavity-nesting colonies distributed broadly throughout California did not detect any colonies with *Varroa*. A second survey in 1993 showed that the feral population had been decimated, especially in areas that had significant commercial beekeeping nearby, reducing the population of feral colonies to 13% of its original size. The empty nest sites contained the smoking gun of destruction, *Varroa* skeletons. Prior to the arrival of *Varroa*, the life expectancy of a feral nest (continual occupancy) was 3.5 years; after its arrival, it was between 6 months and 1 year.

1.3 Parting View

Flowering plants and bees coevolved over millions of years. Bees and flowering plants danced in a dialectical embrace, each changing to fit the challenges and needs of the other with the sole goal of increasing their individual reproductive success. In the process, they transformed the earth to resemble an impressionist painter's canvas splashed with colors taken from the visual spectrum of bees, the artists of the environment. In the next chapter, we explore how honey bees paint their environment.

2

How Bees "Paint" Their
Local Environments

> *The charm of a hike lies in the totality of the impressions gained. When once the goal is reached, one would not wish to overlook one's memories of flowers and of the shapes of trees that no longer are in the field of vision, nor of the many views that opened temporarily before one.*
>
> Karl von Frisch (1967, p. 526)

The canvas of a honey bee colony is dynamic, changing in size, color, and brushstrokes with seasons, days, and even hours. The dynamics of the color pallet are expressed in the dances of returning foragers and the colors of the loads of pollen that adorn their hind legs, the floral pigments. Beekeepers collect pollen from returning foragers and use it to feed protein to bees during times of low availability; it stimulates the production of larvae. There are various devices used to harvest the pollen, all of them involve some kind of screen the returning foragers must pass through as they enter the hive. The openings in the screen are just large enough for the bee to squeak through but not large enough for the bee plus a big pollen load, carried on the outside of the rear legs. The larger loads are knocked off and fall into a catch container. The pollen traps can be mounted in front of the entrance or beneath the bottom box of the hive. If you remove a catch container at the end of the day, you may observe a beautiful arrangement of different-colored pollen, harvested from nature's flowers. Pollen will appear in layers of yellow, purple, white, gray, red, and others. The layers represent changes in the pollen available at different times of the day and different locations where bees are foraging. The foraging locations of the bees change throughout the day. You can observe this phenomenon on a different level by observing the dances of the pollen foragers throughout the day on combs of an observation hive. Bees indicating the same location by the direction and distance conveyed by the dance will have similarly colored pollen; those indicating different locations may have different colors. Colors of loads as well as the directions of dances change throughout the day. Bees show strong fidelity to floral sources, and many studies have shown that less than 5% of pollen foragers carry loads of mixed pollen.

Bees can forage more than 10 kilometers (km) from the nest, though most forage within 1 km. This represents a huge area that can potentially be covered by a colony and potentially a vast array of different patches of flowers and floral species. For example, if you recall your high school geometry, the area of a circle with a radius, r, is equal to πr^2. If bees are foraging up to 10 km around a nest, they're covering an area of 314 km^2. If they're foraging half that distance, up to 5 km, the area covered is nearly 80 km^2, still a very large area. With an observation hive, you can map where a colony is foraging at any given time by using a protractor to determine the angle of the dance from the top of the comb (Figures 1.5 and 1.6), representing the direction of the sun, and a stopwatch to determine the time of the straight portion of the waggle dance. If you plot the dances over the course of a day, you can track how the bees are foraging in different places at different times.

2.1 Colonies as Optimal Foragers

Bees are able to exploit their environments temporally as well as geographically. The results are like brushstrokes on the landscape. Plant species secrete nectar and present pollen at different times of day. Bees tend to forage only during the time of day that their resource is producing. From day to day and hour to hour, the floral patches that bees of a single colony are visiting change their profitability, and colonies are able to adjust their foraging force accordingly. Patches of flowers are ephemeral: they come and go in a matter of hours to weeks. Foragers can outlive their patches and become unemployed or underemployed. It makes no sense for them to remain unemployed when the flowers are no longer there or stop yielding or when the foraging reward is less than the cost to collect it. So, foragers can once again become recruits. As the reward value of a resource declines or exceeds the cost to collect it, a forager makes fewer trips over time, stops foraging, and is then recruited to a new source. This can take place over a period of days.

Each day there are unemployed foragers. These aren't only those foragers whose resource has become unprofitable but also new, naive foragers— bees that have recently matured to become foragers. During the spring and summer, queens can lay 1,500 eggs per day (some may lay more)—that's one per minute. In that case, 1,500 new bees emerge each day and, after about 3 weeks, transition to foraging. Both the unemployed and new foragers are potential recruits. Recruits gather on the dance floor of the nest, an area on combs very near the entrance where returning foragers, roughly 25% of the hive population, conduct their recruitment dances. New recruits attend these dances at random, meaning they go to the dance floor and just bump into dances and get recruited to a particular patch of flowers. The competing patches of flowers, in the foraging range of potentially more than 300 km^2, vary in the value of their nectar and pollen. New recruits are differentially recruited to the patches that are more profitable. This makes sense if colonies are optimal foragers, as we would expect from evolutionary theory. But how do colonies shift their foraging forces from less profitable resources to those that are more profitable? The potential recruits attending dances don't know the relative value of the dances they attend, and no individual dancing bee has any information other than her perception of the reward value of her own patch; so how can this lead to differential recruitment to the more valuable resources?

The recruitment dance itself does not convey any information about the value of a resource relative to another, nor does it convey the value of the resource in an absolute or subjective sense. However, bees foraging on more valuable resources are more likely to perform a dance and will dance for a longer duration. So, for an equal number of bees foraging at each resource, there will be more dances indicating those that are more valuable, and more recruits will go to those resources. If this were the only mechanism to differentially recruit bees to higher-valued resources, it would work very slowly, too slowly to make much of a difference because resources themselves are ephemeral. However, it's been shown repeatedly that colony responses are rapid, the allocation of recruits rapidly shifting to the resources with the highest reward. Something is missing from this explanation.

What is missing is that the dances themselves should reflect relative differences in the foraging profitability of the different patches of resources. How do bees determine the profitability (value) of their resource? Karl von Frisch and his students figured this out for nectar over decades of research. Factors affecting profitability include the following:

1. The sweetness of the sugar solution. Sweeter nectar means more sugar and more energy.
2. The purity of the sweet taste. Contaminated sugar solutions are less valuable to bees.
3. The weight of the load collected. A heavier load means more nectar available and/or a higher sugar concentration.
4. The nearness of the food source. Sources that are closer are more valuable because less energy is expended going to and from the source.
5. Floral fragrance enhances the value of a resource.
6. The shape of a flower. Feeding stations containing sugar solutions are valued more highly if they have shape properties of a flower.
7. A constant flow from a food source over multiple foraging trips increases its value to a forager.
8. Improving food quality makes it even more valuable. The perceived value of a solution offered at a feeding station, say 30% sugar, is more valuable if it has increased from 10% than if it has decreased from 50%. (This is a psychological trait of humans too!)
9. The nutritional status of the colony. If a colony is in a poor nutritional state—little stored honey, not much nectar coming in—then a bee will value a low-quality resource much more and is more likely to dance and to dance for a longer time than if the colony has lots of stored honey and lots of nectar coming in.

How does a bee determine the nutritional status of the colony? She doesn't, at least not directly. But the time it takes her to unload her nectar is an indicator of the nutritional state. Nectar foragers pass their loads to receiver bees at the entrance or near the dance floor. Receiver bees take the loads they collect from foragers and find food storage cells to place it in. If the colony has plenty of stored honey, it takes them longer to find a cell to unload into, decreasing the number of receiver bees available on the dance floor to unload foragers. Or if there are a lot of foragers returning with nectar, a greater number of receiver bees are employed to move the nectar loads from the dance floor to the storage combs. This also decreases the number of receiver bees available to unload individual returning foragers. The time delay decreases the likelihood a forager will dance and the duration of the dance. Only the bees collecting the relatively higher-valued resources, determined by factors 1 to 8, will dance. This is the missing piece of the model that rapidly shifts recruitment to the highest-rewarding resources.

We know much less about how bees value pollen. We don't have a list of factors that determine profitability. However, we do know that unloading time is a factor. Returning pollen foragers unload their pollen into cells located near the brood. They walk around inspecting cells until they find one that's empty or only partially filled. Then they back into the cell and scrape the pollen loads off of their rear legs. The longer it takes them to find a suitable cell, the less likely they are to dance. Time to find an empty cell correlates with how much pollen is stored, an indicator of protein nutritional state. Keith Waddington showed that pollen quality is also an important factor. He used the dance as a method to test how bees report how they evaluate the resource, as did von Frisch and his students. Pollen was presented to bees in open dishes near the hive. The pollen had been collected in pollen traps, dried, and ground into a fine powder. Bees visited the pollen feeders and collected their loads. When Waddington diluted the pollen with a dry inert substance, the bees reduced their round dance probabilities and dance rate, the number of turns per interval of time.

Von Frisch defined *dance threshold* as the concentration of sugar solution sufficient to elicit a dance from a forager. He tested dance thresholds by increasing or decreasing the concentration offered at feeders and noted whether the forager performed a dance. He recognized that dance thresholds aren't fixed; they vary with the season. He remarked that it was difficult to train bees to the feeders during midsummer in Germany when flowers were plentiful and nectar abounded. Bees have a response threshold to sugar that can be demonstrated using the proboscis extension response (PER) assay (Figure 1.4). The response threshold to sugar affects learning. Bees that respond to lower concentrations of sugar (have low response thresholds) perform better on learning tests. It's reasonable to assume that sucrose responsiveness affects how bees subjectively value their resource and consequently affects dance thresholds.

Many studies have shown that the assessment of reward value by bees is subjective rather than absolute. Ricarda Scheiner collected incoming foragers over the course of the flowering season in Berlin, Germany, April through September, and subjected them to the PER assay to determine their response thresholds to sugar. Foragers returning April through May were very responsive. (I should point out here that bees that are very responsive to sugar respond to the lowest concentrations of solution; they have low response thresholds. Bees that are less responsive to the low concentrations [i.e., they respond only to the higher concentrations] have high response thresholds.) Responsiveness was lowest May through July, then high again late July to September, after the main nectar flow. Their responses to the solutions weren't constant; they changed with the season. Learning tests were also performed on the bees. Bees performed better during the times of the season when they were the most responsive to sugar, corresponding to the observation of von Frisch. Learning performance was subjective; it correlated with their perception of the reward.

The concentrations of sugar solutions fed to bees change their perception and thereby their response thresholds to sugar. When they're fed a high concentration, they have higher response thresholds (are less responsive) than when fed less concentrated solutions. They adjust the way they value the sugar based on what they've been getting. If you feed one group of bees a higher concentration and a different group a lower concentration and then test their response thresholds, they'll be very different. But if you then feed both groups the same concentration for 1 day and retest them, they'll be the same. Differences are dynamic, based on recent experiences.

One can think of a colony as a communal stomach. Incoming nectar foragers pass their loads to receiver bees, who in turn share it with other bees before storing some of it in cells on honeycombs. Those bees in turn share it with others, etc. Studies using radioactively labeled sugar have shown how sugar solution is rapidly distributed throughout the colony; soon, all individuals have the labeled sugar. Tanya Pankiw demonstrated how the communal stomach modulates sugar responsiveness by placing individual colonies inside large flight cages. She offered sugar solution at a feeder inside the cage, and bees actively foraged and collected the solution. After the bees had foraged for a while, she collected nest bees from the brood combs, bees that hadn't foraged, and tested their response thresholds. When foragers brought in solutions with high sugar concentration, the nest bees had higher response thresholds, were less responsive, than when foragers collected less concentrated solutions. The response thresholds of the whole colony were modulated by the concentration of incoming food. This represents global information about the quality of the food collected that may be used by foragers, perhaps affecting their dance thresholds.

2.2 Individual Bees as Optimal Foragers

Honey bee foraging behavior is shaped by natural selection on two levels: (1) the colony, where the allocation of the foraging workforce shifts between floral patches, concentrating the workforce of roughly 10,000 bees during the spring and summer on those patches that are more profitable, and (2) the foraging behavior of individual bees to maximize their foraging efficiency. Every nectar foraging trip by every bee involves costs and benefits in terms of energy and time expended and energy collected from the flowers in the form of sugar from the nectar. When a bee leaves the nest, she carries with her about 2 microliters (μl) of honey that will supply her foraging energy. The amount varies with how far she must fly. She expends time and energy flying to the source. When she gets to the flower patch, she expends time and energy collecting nectar and flying between flowers. She gains the energy provided in the nectar collected. The larger (heavier) the load she collects, the more energy she expends flying between flowers. Also, the more she exploits a patch, the longer it takes and the more energy it takes to find a flower that hasn't already been visited and still yields nectar. You can imagine that at some point she could end up expending all the energy she gains in one flower seeking a reward in the next flower. She certainly should return to the nest before that occurs. But when?

The decision to return to the nest is based on the forager's patch departure rule, which in turn is based on the bees' loading curve (Figure 2.1), concepts borrowed from economic theory. In general, the longer she remains in the patch, the less energy she gains per flower visited or time interval spent foraging. This can be due to the depletion of the patch from foraging activities or increasing flight energy expended as the crop weight increases. If the loading curve is linear, there's no diminishing gain for each additional time interval; then, the optimal patch departure is when the bee is fully loaded with about 60 μl of nectar. However, many studies going back to von Frisch have shown that bees don't normally fill their crops, a result that's predicted by mathematical optimality models. In general, optimally foraging bees should visit fewer flowers (return with smaller loads) if they must travel a shorter distance to the patch, they expend more energy in the patch finding rewarding

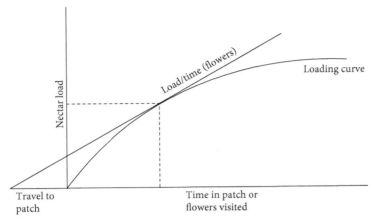

Figure 2.1 Patch departure decisions of foragers. The diagram illustrates the important elements involved in the decision of an optimally foraging honey bee to return to the nest with the load of nectar she has collected. Why not just fill up? Across the bottom (*x*-axis) the time scale is divided into the time to travel to the flower patch where she forages and the time she spends in the patch collecting nectar. The vertical line on the left (*y*-axis) is the total amount of nectar she has collected. The loading curve line represents the cumulative load she can collect in the patch. It isn't a straight line because the longer she stays in the patch, the less she collects for each subsequent unit of time. For example, as her crop fills, it takes her longer to add more to it because of the back pressure of full crop. Or as she visits flowers the flowers become depleted, and it takes her more time to find one that she hasn't already harvested. (Imagine going to the grocery store without a basket. You need to get 10 cans of something. When you start taking them off the shelf, you can easily hold them, and you collect them quickly. But as your hands and arms fill, each one takes you longer to fetch and find a way to carry.) The optimal solution lies at the intersection of the straight line (tangent) and the loading curve line. It represents the point, in terms of time or flowers visited, at which she should return to the nest, unload, and start over. If she stays longer, she will actually collect less because she becomes less efficient.

flowers, each flower yields less nectar, or the sugar concentration of the nectar is reduced.

The "currency" being optimized by bees could be the net energy gained per unit of time or foraging efficiency. The net gain per unit of time is represented by (gain – cost)/time. In this case it's important to get as much food as possible back to the nest in as short a time as possible, a strategy that birds feeding hungry nestlings employ. Empirical studies, however, have shown that honey bees maximize efficiency rather than net gain/time. Efficiency is (gain – cost)/time. Honey bees have a limited number of miles they can fly. They can't be fully replenished for subsequent flights. They're limited by the glycogen stored in the flight muscles, a major source of flight energy. The glycogen stores apparently aren't readily restored. Bees are also limited by wear and tear on the wings. If you look at the wings of an older forager, you'll notice that they have pieces missing—they wear out. A foraging honey bee is like a car that only gets one tank of gas. If you drive it fast, your gas mileage decreases, and you will go a shorter distance but get there faster. If you drive it more slowly, you get better gas mileage and will travel farther—you're driving more efficiently.

2.3 Pollen Foraging

The diversity of the floral environment would be reduced if only the highest-valued resource was visited by bees. The landscape canvas would become monotonous, covered with fewer and fewer floral species with greater and greater patch sizes. However, honey bees are constantly finding and exploring new resources and set a higher subjective value on rarer plants providing nectar and pollen, a premium for diversity. Much is known about how bees subjectively evaluate nectar because nectar has an easy absolute value to assess, the amount of energy contained in the sugar. Much less is known about the reward value of pollen other than that some pollens provide better nutrition than others. Interestingly, there's little evidence that bees assess pollen value based on nutrition, probably because the valuable protein is locked inside a hard shell with a pore that must be digested by the bees to unlock the reward.

Foraging honey bees are often grouped into pollen or nectar specialists. However, many or most foragers collect both pollen and nectar. Some bias their foraging toward collecting nectar, others toward pollen. A bee with a strong nectar foraging bias accidentally brushes against the anthers of a flower and gets pollen on her body hairs while she probes the nectary. After collecting nectar, she grooms herself and stores a small amount of pollen on her hind legs. When she returns to the nest, she brings a load of nectar and a small amount of pollen. Other bees work the flowers and collect the pollen.

Their behavior is different. They land on or near the anthers, wallowing and getting large amounts of pollen adhering to their branched body hairs (Figure 1.2). In the process, they may collect some nectar that gives them the energy they need to continue foraging, or perhaps they inadvertently taste the nectar with the taste sensors on the tarsi of their feet and reflexively feed on the nectar. When they return to the nest, they'll have large pellets of pollen on their hind legs and relatively little nectar in their honey stomachs.

Pollen and nectar collecting are intimately related because of loading limits experienced by foragers. A fully loaded bee collecting only nectar can carry about 60 milligrams (mg). A fully loaded pollen forager carries about 30 mg. The lower load weight for pollen is probably because the aerodynamics of large pollen balls on the outside of the hind legs reduces flight efficiency. If you sit at the entrance of a hive and collect returning foragers and measure their loads, you will get a distribution like that shown in Figure 2.2. What is striking is the scatter of data points throughout the lower diagonal half of the figure. Many bees returned with only nectar, others with only pollen; but a very large number collected both. There's also an obvious constraint placed on bees—defined by a line running between the maximum loads possible for pollen and nectar. Each of these bees made loading decisions about collecting pollen and nectar while they visited flowers. Loading decisions are based on the integration of information obtained in the nest and while foraging in the patch modified by the genetic makeup of the bee.

To our aesthetic eye, bees are artists painting the environment for our appreciation. But to the complex web of life that depends on the plants for shelter, food, or nesting materials, they're environmental engineers. They engineer the environment as they paint it with their pollination activities. They also engineer their own nest environment. The architectural blueprint is in their genes.

The nest has a specific architecture that determines the location of the brood and the honey and pollen stores. Empty honey storage space stimulates the collection of nectar. I have discussed the regulation of nectar foraging. Larvae stimulate pollen collecting, while stored pollen inhibits it. If you remove larvae or increase the combs of stored pollen, the foraging force of thousands of bees will shift toward more nectar collecting. If you add larvae or remove pollen, the foraging force shifts to more pollen collection. You can observe this by watching the entrance and seeing the number of bees returning with pollen. Low stored pollen and high numbers of young larvae also result in an increase in the diversity of pollen coming in, demonstrated

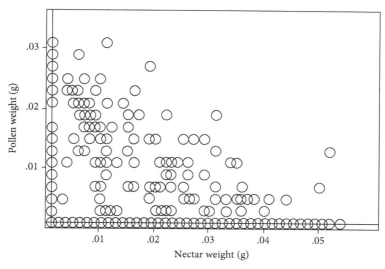

Figure 2.2 Pollen and nectar loads of bees. The diagram shows the loads collected from a large number of returning foragers. The bees were collected at the entrance of a hive, and their pollen and nectar loads were weighed. Each dot represents the load of one bee. Some bees collected only nectar, others only pollen. Many bees collected both. The maximum nectar load was about 60 mg and the maximum load of pollen, 30 mg. Pollen loads are carried on the outside of the legs and probably introduce aerodynamic drag to the bee, resulting in a higher energetic cost to transport. Nectar is carried inside the body in the crop, or nectar stomach, and doesn't introduce aerodynamic drag. The scatter of points reflects the availability of pollen and nectar for each individual bee on her foraging trip. Reprinted with permission from *The Spirit of the Hive: The Mechanisms of Social Evolution*, by R. E. Page, Jr., 2013, Harvard University Press. Copyright © 2013 by the President and Fellows of Harvard College.

by the different colors of pollen loads collected. Pollen diversity increase is probably due to the increased acceptance of lower-quality pollen, a consequence of lower response thresholds for pollen quality stimuli. The increase in pollen collecting will continue until the quantity of stored pollen meets the needs of the colony.

Each bee adjusts her loading decisions as she forages based on her individual perception of the colony need (stored pollen and larvae to feed) and the floral resources (Figure 2.3). She assesses the colony need after she returns from a foraging trip by walking along the boundary between the

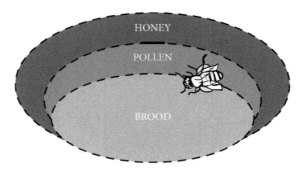

Figure 2.3 Cartoon of a returning pollen forager. The loaded bee finds a comb that contains brood and pollen. She walks along the boundary between them in search of a cell that can contain her load. During this search she's exposed to two of the major pollen foraging stimuli, larvae, a positive stimulus, and pollen, an inhibitor. Pollen cells are loaded to about two-thirds full, while honey-containing cells are filled to the top. The pollen forager must back into the cell and scrape off her pellets by rubbing her legs together. Pellets are packed into the cells by bees pressing them with their heads. Bees that put nectar into cells fill it with their proboscis, which can be used to top off the cell.

brood and the stored pollen. Nurse bees eat pollen from cells close to the larvae, process the pollen protein into protein-rich secretions, brood food secreted from glands in the head, and feed it to the larvae. If there are many larvae to feed and little incoming pollen, then there will be empty cells at the interface of the brood and stored pollen and a pollen forager will be able to quickly offload her pollen pellets. If there's a lot of stored pollen and not many larvae to feed, then it will take her longer to unload. The probability that she'll do a recruitment dance is correlated with the time it takes her to unload and perhaps affects her foraging decisions. Pollen stored away from the brood also affects foraging loading decisions, but we don't know how, except that the foragers must be able to have contact with the stored pollen for it to have an effect.

Larvae manipulate the physiology and behavior of adult workers. Larvae produce chemicals called *pheromones*. They create an odor environment that has immediate and delayed effects on the behavior of bees. They stimulate nurse bees to feed the larvae. They directly affect pollen-collecting behavior by "releasing" that behavior in foragers. Releaser effects of pheromones are like triggers of preprogrammed behavioral patterns, releasing the neuromotor responses that generate behavior. But the odor environment produced by larvae, especially young larvae, also "primes" (prepares) young worker bees to become pollen foragers when they initiate foraging during their third or fourth week of life. It alters their loading decisions.

Colony need is in the eyes of the beholder. Each bee has a different "target" for the amount of stored pollen needed. The difference in targets among foragers is at least in part a consequence of their genetic makeup. Honey bee queens mate with many males, some estimate in excess of 20. They mix the sperm of their many mates and use them throughout their egg-laying lives of 1 or 2 years. The workers in a colony represent a tremendous diversity of genetic composition derived from the many fathers and the mixing of the queen's genes into the eggs she produces. Every one of her several hundred thousand eggs laid during her life is different. Honey bee behavioral genes are variable. Many genes exist in different forms with different effects and different levels of gene expression (the complementary copies of RNA made from the genes that will be translated into proteins) or are expressed differently in different tissues. This generates genetic variation in behavior. Bees with different genotypes (the collection of their variable genes) make different loading decisions. This has been demonstrated repeatedly in many different ways but especially as a consequence of a 22-year breeding program where bees were selected for how they collect and store nectar and pollen. Two strains were produced that differed dramatically in their sugar response thresholds, foraging decisions, perception of pollen quality, and scouting behavior.

We do have a window into the information processing of the bee through which we can take a glimpse at loading decisions. The PER assay can be used to determine the sugar response threshold of bees. Response thresholds to sugar are in part genetically determined. Bees from different genetic sources have predictably different average response thresholds. Differences exist even in young adult bees that are less than 4 hours old. These differences persist and affect their foraging decisions. Very young bees that respond to the lowest concentrations of sugar are more likely to bias their foraging toward collecting pollen and water. Bees with nectar-collecting biases do not respond to the lowest concentrations; they have higher response thresholds. Some bees only respond to very high concentrations of sugar solution and are more likely to return from foraging trips empty. It's possible that these bees are "scouts" that collect from and recruit only to the floral resources with the highest sugar rewards. Response threshold assays of returning foragers confirm these results. Returning pollen foragers have lower sugar response thresholds than returning nectar foragers.

The dance floor of the nest is an arena where foragers compete to recruit currently unemployed, experienced foragers and brand new naive bees that are coming of age. It's like the booths and displays of a bazaar, each forager hawking her goods. Potential recruits bump into dances at random as they patrol the arena, but the dances that get the attention aren't random. Bees with a genetic disposition to collect more pollen follow more of the dances of pollen foragers, getting information about the most attractive sources of protein. Bees that were selected for increased pollen foraging also are more likely to return from a scouting trip with a load of pollen. The pollen foragers are the Monets of the colony and the chief engineers of the environment.

2.4 Parting View

In these first two chapters, we viewed the evolution of bees as a group, all 25,000 species of them, as painters of the earth as they coevolved with flowering plants. The local impact of bees on the environment is inescapable when observing patches of spring wild flowers. The specific impact of honey bees is obvious when viewing agricultural cultivations or when enjoying our meals. We viewed more closely the effects of honey bees on cultivated and uncultivated landscapes in geographical and temporal scales and the anatomical and behavioral adaptations that have made them so successful in exploiting rich floral resources. I've stressed the aesthetic effects of their activities in this chapter and the behavioral mechanisms they've evolved with which they dynamically exploit an ever-changing resource environment. But one could take a different view, one of honey bees as environmental engineers, constructing environments that benefit themselves as well as many other species that depend on and interact within those environments. That is the view in the next chapter.

3

Environmental Engineering

> *And NUH is the letter I use to spell Nutches*
> *Who live in small caves, known as Nitches, for hutches.*
> *These Nutches have troubles, the biggest of which is*
> *The fact there are many more Nutches than Nitches.*
> *Each Nutch in a Nitch knows that some other Nutch*
> *Would like to move into his Nitch very much.*
> *So each Nutch in a Nitch has to watch that small Nitch*
> *Or Nutches who haven't got Nitches will snitch.*
>
> Dr. Seuss, *On Beyond Zebra* (1955)

My favorite book by Dr. Seuss is *On Beyond Zebra*. In the book, young Conrad Cornelius o'Donald o'Dell is learning from his friend that there's another complete alphabet beyond "z," an alphabet needed to spell the names of strange, unknown animals. True to Dr. Seuss' style, the book is written as much for the parents as for the children to whom they read. Of the letters beyond "z" and the animals attached to them, my favorite is the letter "Nuh." As a biologist, I recognize that Seuss is familiar with the niche concept (he uses *Nitch*, a common misspelling of *niche*) in ecology, and his poem and the corresponding picture give a good description of the concept at the time.

The concept of an ecological niche dates back to the turn of the 20th century. The term refers to a recess in a wall where statues are displayed. In the Seuss caricature of a niche, small, strange-looking animals sit individually within their own niche in the side of a mountain, every niche occupied, every Nutch looking worried that its niche will be taken. The ecological-niche concept was used first in a research paper by Joseph Grinnell in 1917. He defined a niche as the collection of properties of the environment in which a species forages and exploits resources. To him it was much like the Seuss cartoon, a recess in a wall that individuals of species occupy. Species adapt to the niches they occupy by evolving behavior and anatomical features that help them better exploit it, a view of organisms as modeling clay, evolving necessary traits for survival and reproduction, the environment being the sculptor.

In 1927, Charles Sutherland Elton added to the niche concept by expanding it to include the effects of other species sharing the environment and most importantly the effects of an individual organism on its own environment. Now organisms that occupy niches are part of the construction of the niche. G. Evelyn Hutchinson expanded the concept in 1959 into what's now most widely used in ecology and defined a niche as an "n-dimensional hypervolume." This is the concept I learned when I took ecology in 1973. What this means is that you specify the organism within its environment by its *niche vectors*, all of the individual elements of its environment that define it as belonging to a given species. For example, this could be the available food and shelter, the species' predators and competitors, etc. Over time, the population (species) becomes adapted to its position in this hypervolume of environmental factors. This definition diminished the effects of the organism itself on its own environment, a modeling clay–sculptor concept. Recently, David Krakauer and associates co-opted the Hutchinson concept and refocused on the importance of the organism in constructing its own niche.

3.1 Engineering the Environment

Honey bees are environmental engineers. Through their activities they affect the niches in which they live. When those activities affect the environment in ways that affect other species, it's called *environmental engineering*. When the activities affect the environment (niche) of the organisms engaged in them in ways that affect their own survival and reproduction or their descendants, it's called *niche construction*. It's obvious that the pollination activities of honey bees have broad effects on other organisms that use those plants. Some animals eat the plants or their seeds and fruits; others use them for shelter, and the plants decompose and can be used as food for detritus feeders, to increase the fertility of the soil for other plants, etc. Many species of solitary and social bees also use the flower resources resulting from honey bee pollination. Remember, there are more than 25,000 species of bees, most of them leading solitary lives and make a living from flowering plants. It's not so clear that they benefit directly from their own niche construction pollination activities. Are they exploiting the resource, or through their own activities are they growing seeds for new plants from which they may ultimately benefit? The foraging activities of all bees were responsible for the adaptive radiation of bees and flowering plants, and that changed the face of our planet.

We distinguish between niche construction strategies to protect organisms from the uncertainties of the environment by insulating themselves from it and strategies that change the external environment to be more beneficial. We further distinguish between niche construction strategies that reap direct benefits to those that currently occupy the niche from indirect effects that benefit future generations or other close relatives.

Strategies to Insulate Colonies from the External Environment

The nest environment of the honey bee was engineered by natural selection over millions of years and is constructed by the activities of thousands of individual bees. One can imagine that the first step in the evolution of the protective nest of the honey bee took place more than 65 million years ago, when the ancestor of the honey bee made the transition from a solitary to a primitively social state. Female solitary bees form nests, often by hollowing out twigs or by burrowing in the ground or hardwood. The female constructs a cell that she provides with a loaf or ball of pollen mixed with nectar, lays an egg on the pollen, and closes the cell. This is a common method of nesting in the non-social bees. The larva hatches from its egg, consumes the pollen, pupates, emerges as an adult, then leaves the nest and disperses. After dispersing, males and females mate; and the mated females either find shelter for the winter or found a new nest, depending on when they emerged and the length of the breeding season. Males typically don't engage in nest construction, nor do they provision the nest. Common in the animal world, their only contribution to their offspring is sperm.

The first transition from solitary to social required that the bees from a nest not disperse but remain together. In some solitary relatives of the honey bee, daughters remain in the maternal nest for some time before dispersing. In some cases, they help their mother raise more brothers and sisters; in others, one may take over the maternal nest and raise her own offspring. In some carpenter bees, females leave the maternal nest; but sisters form groups and jointly construct chambers in the ends of pithy plant stems, where they overwinter, then disperse in the spring and form their own nests. Sometimes, sisters will form an alliance, a primitive social group, and jointly found a nest and divide up the jobs of nest construction, egg laying, guarding the entrance, and foraging. Accepting another female in one's nest is a big first step toward sociality. Individual females protect their nests from intruders that often try to take over a nest that's already been constructed, saving them the effort of initiating their own. A female must learn to recognize who belongs and who's an intruder. Hence, nestmate recognition was an important early step in the evolution of the nest environment.

Primitive nest sites are ephemeral; each nest lasts just one season. Such seasonal nests are found in social bumble bees. Each spring a large, mated queen founds a new nest, often an abandoned mouse den. She excavates, lays fertile eggs that will become female workers, and forages for pollen and nectar to provide for developing larvae. When the first batch of eggs (typically 12 to 16) becomes adult worker bees, she ceases to forage and remains in the nest. The first workers are tiny compared to the queen, but they work themselves to death foraging and caring for the new brood that results in female workers that are larger in size than the first batch. This continues throughout the spring and summer until the final brood is produced. These are the large reproductive males and females that will constitute the next generation. The reproductive bees leave the nest, queens mate, males die, and the mated queens find refugia to survive the winter and start the cycle over. The residual workers from the maternal nest slowly die off until the maternal nest is vacant. This is a typical life cycle of many social bees and wasps. The colony and the nest site survive just 1 year, with reproductive males and females being produced in relatively large numbers at the end of the season; it's an annual life cycle.

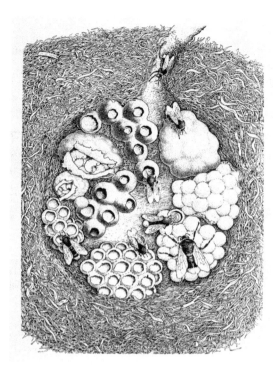

Figure 3.1 Drawing of a bumble bee nest. From *The Insect Societies* (fig. 5.9, drawing by Sarah Landry), by E. O. Wilson, 1971, Belknap Press of Harvard University Press. Copyright © 1971 by the President and Fellows of Harvard College.

It's likely that the progenitors of honey bees had an annual life cycle like that of bumble bees (Figure 3.1). But something happened. Honey bee queens live longer than one season; they're perennial and don't produce a burst of reproductive males and females at the end of the season. The life of a colony may be measured in terms of how long a succession of mothers and daughters continuously occupy the nesting site. Old nest sites are very attractive to a colony seeking a new nest site. Nest sites, once abandoned, are frequently reoccupied by swarms, so it's difficult to determine how long a colony exists. Honey bees have an orderly succession of passing the nest site from mother to daughter, either through *swarming*, where the mother queen leaves the nest with an entourage of about half of the workers, or by *supersedure*, where the old queen relinquishes the nest and colony to a daughter queen, then dies.

The nest became a fortress to defend, a massive nursery to raise new bees and a processing plant and storehouse for food reserves to keep the colony alive during times of food scarcity. It's been characterized as a factory within a fortress. Over time, bees evolved and engineered the nest to best meet those needs: selection of a secure, defensible nest and construction of wax comb that became the nursery, food repository, and social substrate. Society was engineered to provide the workforce to care for large numbers of developing bees—gather, process, and store surplus honey and nectar, thermoregulate the nest, raise new bees under extreme temperatures, and engage in hygienic behavior to reduce the pathogen and parasite loads that come with dense populations and long-term habitation of a nest.

Engineering the Nest

The fortress of the honey bee has design elements that are selected from pre-existing structures in the environment and others constructed by the workers. It's driven by three needs: (1) it must be defendable, (2) it must be sufficiently insulated to regulate the temperature, and (3) it must accommodate the structural needs of the factory, the combs for storing food and rearing bees. Bees select a nest site by looking among existing nest site options using a set of criteria that are probably driven by simple rules. The process of site selection and internal construction involves socially engineered behavior.

The Swarm

Colonies seek new nesting sites when they undergo reproductive colony fission and when they abscond, abandoning their home nest in a search of a better location. When a colony undergoes reproductive fission, a new virgin queen is produced that will inherit the old nest, and the mother queen leaves with roughly half of the workers and seeks a new nest site. The process of reproductive swarming is coordinated, occurring at a time of year, the spring, when both parts, the old nest with a new queen and the new nest with the old queen, can successfully produce a new family of thousands of workers and bring in resources to survive the winter. But to survive, the homeless cloud of bees must find a suitable habitat to settle.

Bees prepare to swarm when the combs are full of honey and brood and there are few open cells for the queen to lay eggs. This usually corresponds to an abundance of younger workers that are hanging around on the combs in the brood nest with nothing to do, congesting the brood nest and getting in the way. This usually occurs in the spring when the colony is growing rapidly in worker population as a result of abundant pollen and nectar coming into the nest and the vigorous egg-laying activity of the queen following her winter pause. The queen at this time becomes less active, not moving around as much and covering the combs with pheromones from her feet called *footprint pheromones*. These pheromones inhibit the workers from constructing new queen cells and raising queens. Also, with all of the congestion in the nest and the restricted movement of the queen, the main pheromone she produces that inhibits queen rearing, the queen mandibular pheromone (QMP), isn't distributed efficiently. The QMP is produced in large quantities by paired mandibular glands, distributed throughout the nest by mouth-to-mouth food exchange, *trophallaxis*, and has many behavioral and physiological effects on workers in addition to inhibiting them from raising new queens. In the absence of the inhibition, workers will start preparing the special royal cells in which the new queens will be raised. The cells start as small cups, just nubs of the larger peanut-shaped cells that hang vertically at the bottoms of the combs (Figure 3.2). The queen lays a single egg in each cell that will hatch 3 days later. The new larvae will be fed by the nurse bees an unlimited supply of royal jelly for 5 to 6 days, while the workers keep elongating the cell to give the larva room to grow, then cap the cell. The new adult queens will be ready to emerge about 8 days later. The nurse bees will typically raise between 15 and 25 new queens, far more than will actually be needed or survive.

The status of the queen quickly changes once the workers initiate new queen cells. She's put on a restrictive diet, receiving far less food from the workers. Queens are fed large amounts of royal jelly when they're actively laying eggs. The royal jelly replenishes their nutritional needs and is the source of proteins and fat used to make eggs. An actively laying queen is full of eggs and body fat and usually incapable of flying. The queen must slim down and be able to fly with the prime swarm that will issue from the nest in about 1 week. The workers abuse her, push her around, bite her with their mandibles, and pull on her, making her walk and move, an enforced fitness camp. She slims down rapidly, losing about 25% of her body weight, and is ready to take flight within a few days. The workers hang in festoons on the bottoms of combs, underemployed, covering the combs with a multilayer blanket of bees, even outside the hive's entrance. The quiescent workers are engorged with honey, their honey stomachs packed with about 36 milligrams (mg) of honey, to take on the trip to their new home. While they hang in their engorged, replete, lethargic state, their bodies are busy making wax scales

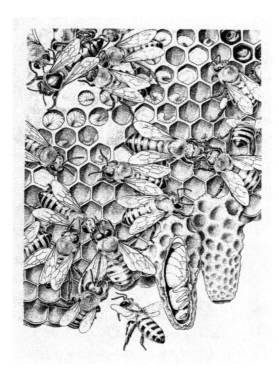

Figure 3.2 Drawing of part of a comb from a natural honey bee nest. The hexagonal cells are being used to raise larvae. The two large peanut-shaped cells contain queens. A male bee (drone) is shown at the bottom center of the comb, distinguished by his large eyes that converge at the top of his head. An adult queen is shown in the top left corner surrounded by an entourage of workers. From *The Insect Societies* (fig. 5.16, drawing by Sarah Landry), by E. O. Wilson, 1971, Belknap Press of Harvard University Press. Copyright © 1971 by the President and Fellows of Harvard College.

from four pairs of wax glands located on the underside of their abdomen. These scales will be needed when they find a new home. Other bees, a small cohort of a few dozen, continue to actively scout the surrounding landscape looking for potential new nest sites. These scouts are important for the coordination of the big event to come.

About 8 days after the queen laid the eggs, the new queen cells are about to be capped. The scouts start running through the nest, exciting the inactive workers by getting close to them and vibrating their flight wings, generating a sound called *worker piping*. Slowly the bees wake up and heat up their bodies.

The scouts continue to push through the mass of waking bees, buzzing their wing muscles in repeated buzz-runs. Others bite and pull the queen, moving her toward the entrance. Soon the level of excitement and movement hits a crescendo, and bees pour out of the entrance, taking flight. The queen is swept out and up in the mass exodus; the swarm is formed.

The first swarm to leave is called the *prime swarm*. It typically consists of about 16,000 bees, 70% of which are less than 10 days old, a young bee's venture. Other swarms (usually only one or two) may issue from the nest over the next couple of weeks as virgin queens emerge and take flight with decreasing numbers of workers. The first after-swarm typically has about 11,500 bees and the second and third about 4,000 bees, with little chance to survive. The swarm forms a cloud about 10 meters (m) in diameter and 3 m in height. The casual observer sees chaos, but in reality it's organized, moving in circles with scouts flying through the swarm directing it while it travels at about 11 kilometers (km) per hour. They don't go far. A few bees land on a nearby branch, expose their scent glands located near the tip of the abdomen, fan their wings, and blow air over the sponge-like gland spreading a chemical scent through the air. The chemical smells like citrus and attracts the other bees flying in the swarm to land on the branch. The queen alights, and soon the swarm is hanging, locked together foot to foot and foot to body in a football-shaped mass, the *swarm cluster*. This is an intermediate rendezvous point, a place for a temporary bivouac from which the scouts can operate and find a new, permanent home. Some bees initiate foraging, bringing back supplies to feed the masses, while the scouts branch out and scour the landscape looking for the perfect nest site. The scouts are part hunter, part surveyor, and part engineer.

Selecting a New Nest
When a scout discovers a potential new home, preferably a hollow cavity in a tree, she goes through an elaborate inspection process taking up to 1 hour. She checks the integrity of the tree—she doesn't want too much wood rot weakening the walls; this home needs to last a while. She also inspects the size of the entrance and its location within the cavity. She doesn't want it too large because it will be harder to maintain the temperature inside and to defend, like leaving a large door open. She doesn't want it too small because the cavity will be too moist and stuffy with the accumulated water vapor of the respiring bees in the winter and the evaporating nectar being transformed into honey in the spring and too hot and difficult to thermoregulate in the

summer. The entrance near the bottom aids in air circulation, allowing the colder and moister air to escape, and reduces the cold winter winds blowing on the brood nest. A south-facing entrance is an added benefit in the northern hemisphere with the advantage of direct sun on the entrance in the winter and early mornings, warming the interior. The elevation of the nest above ground is important as well; it needs to be high and out of the reach of most predators. A survey of the inside of the cavity determines the total volume of the space. Without the benefit of GPS, theodolite, plumb bob, level, or even a measuring stick, she paces the cavity, covering all of the walls and determining the vertical and horizontal dimensions from which she uses her brain—no slide rule, no calculator, and no digital computer—to calculate the volume. The total process may require several trips, flying back and forth to the cluster of her homeless, waiting sisters.

Other scouts find other potential nest sites, each doing an elaborate evaluation of her own. When they return to the swarm, each performs a dance indicating the distance and direction to the prospective home they found and inspected. The dance probability and duration of the dance are dependent on how excited they are about their own find. New scouts inspect nest sites based on dances they attended, return and dance, or not, based on how they evaluate it as a new home. Altogether as many as 500 scouts, former experienced foragers, might be involved, scouring 70 km^2 of the landscape, searching for the best site. Each returns from making her assessment and dances, competing for the attention of other scouts to examine what has been found. This may go on for hours or several days, but when the colony reaches a consensus on which nest will make the best home, the cluster disintegrates, breaking up into a rotating cloud of bees, and moves toward the new nest site led by the scouts streaking through the swarm and showing the direction. When they approach the site, a scout alights on the entrance and exposes her orientation scent gland signaling the location. She's quickly joined by others, marking the path to the entrance to the new home. Soon the queen arrives and enters, followed by the migrant workers, occupying the space.

In the end, the bees select a home as close as possible to the following properties among the discoverable prospects: a cavity of about 40 liters containing combs from previous honey bee colony occupants sitting high in a tree that's sturdy, not necessarily alive, with a south-facing entrance 12.5 to 75 square centimeters positioned near the bottom of the cavity. Once the home has been found and occupied, construction begins.

Nest Construction

Like any big renovation project, first the wrecking crew comes in. They clean out the debris in the cavity and plane down the walls using their mandibles, excising rotten wood that might still contain the live fungus that hollowed out the tree. They don't want it working to deteriorate the inner walls and weaken the structure. They coat all the interior surfaces with a resinous material, propolis, obtained from plants; and all cracks and extra holes are plugged. The exterior of the entrance is also planed smooth and coated with propolis, for a still unknown purpose. Propolis may have antibiotic and anti-fungal properties and may prevent the growth of fungus. While the cavity is being sealed off from the outside, a construction team busily builds the wax comb infrastructure.

The wax comb builders are younger than the scouts and have been pre-paring for the construction project for several days by gorging on honey and sitting around and doing little besides making wax scales. The comb is engineered to meet a design imperative to maximize strength while mini-mizing construction materials. Wax is costly, each kilogram (kg) costing about 6 to 9 kg of honey. Every kilogram of wax will produce about 77,000 worker cells, requiring the secretion of about 990,000 scales. A typical 40- to 45-liter nest will consist of about 100,000 cells (drone and worker size) made from about 1.3 kg of wax, or about 1.3 million wax scales. The finished product is capable of storing 22 kg of honey for every 1 kg of wax, a phenom-enal ratio of materials to storage weight of 22:1.

To achieve this very efficient 22:1 ratio, bees construct comb that hangs vertically and is double-sided with hexagonal-shaped cells on the op-posing parallel surfaces, shared walls with adjacent cells, and shared bases with the cells on the opposite side of the comb. Cell wall thickness is 0.073 millimeters. The hexagonal shape is optimal because there's no wasted space between cells and it provides the most cells per area of comb. In addition, the cells are nearly cylindrical in shape, providing a chamber that accommodates the shape of developing larvae and pupae. Charles Darwin in the *Origin of Species* (1998 Modern Library, paperback edition) gave a clear description of the blueprint for comb construction and described the calculus used for the construction protocols, both engineered by natural selection:

> a score of individuals work even at the commencement of the first cell. I was able practically to show this fact, by covering the edges of the hexagonal wall of a single cell, or the extreme margin of the circumferential rim of a

growing comb, with an extremely thin layer of melted vermilion wax; and I invariably found that the colour was most delicately diffused by the bees— as delicately as a painter could have done it with his brush—by atoms of the coloured wax having been placed, and worked into the growing edges of the cells all round. The work of construction seems to be a sort of balance struck between many bees, all instinctively standing at the same relative distance from each other, all trying to sweep equal spheres and then building up, or leaving ungnawed, the planes of intersection between these spheres. (pp. 346–347)

Darwin continues his description and ends with "the comb of the hive-bee, as far as we can see, is absolutely perfect in economising labour and wax" (p. 349). Darwin went on to show how bees could build to these specifications in the dark hive without a forewoman watching over them, without a set of blueprints to consult, without a level, plumb bob, or measuring tape.

The bees begin by attaching wax to the top of the nest cavity, laying the foundation for new combs (Figure 3.3). The propolis coating provides a smooth, secure substrate on which to build. Wax is taken from the glands, the wax mirrors of the four terminal segments on the underside of the abdomen, with a spine located on the tarsus of the hind leg. It's then moved to the mouthparts, where the wax has liquid added from glands in the head and is chewed, manipulated, and shaped. It takes a bee about 4 minutes to manipulate and deposit a wax scale. Hundreds of bees engage in the construction of a single cell, but an individual may be actively working on a cell for only 30 seconds. On close examination, the construction seems chaotic. One bee may put a piece of comb in place, and another comes along and removes it and uses it somewhere else. But with thousands of bees working, the net result is progress. The cell bases are built first, and then the walls of the cells are extended upward at a 13-degree angle. The upward slope keeps the nectar and honey from running out and the larvae from falling from their uncapped cells. Larvae become restless when hungry and crawl out of their cells, a fatal consequence because there's no way to put them back in. The cells near the top of the comb, where they attach to the top of the nest, are thicker than those farther down, adding strength to the cells supporting the most weight of the comb. When finished, an average cavity will be filled with eight combs, about 2.5 m^2, hanging parallel to each other, attached at the top and sides of the cavity but not attached to the bottom, and with a space of

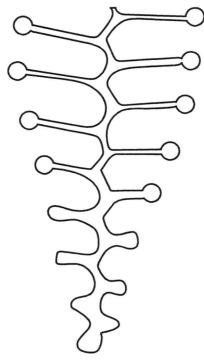

Figure 3.3 Diagram of a comb under construction shown from the side. The center line where the cells from the two sides of the comb join is the midrib of the comb. The midrib consists of the shared bases of the cells. As cells are constructed, the workers build up the top of the cell with excess wax and then extend and shape the cell. From *The Behavior and Social Life of Honeybees* (fig. 50, p. 199), by C. R. Ribbands, 1953, Bee Research Association. Reproduced by permission of the International Bee Research Association (ibra.org.uk).

about 0.95 centimeters between them. Queen cells are constructed along the bottom edges of the combs when colonies are preparing to swarm or replace their aging queen.

Comb for the new nest must be constructed rapidly. The queen needs cells in which to lay eggs to replace the workers that swarmed with her. Their life expectancy of 30 to 60 days in the spring means that their numbers are declining rapidly. Assuming that the colony finds a new nest the same day they swarm and that the workers build comb and the queen begins to lay that same day, it will be 21 days before any new bees emerge as adults. By then, the colony population could be approaching half of its original size. And then, only as many bees can emerge as there were cells produced on the first day of occupying the nest site. The construction workers answer the urgent call and construct 90% of the total comb in the nest in about 45 days.

The nest itself has an organizational structure that's designed to fit the social structure of the colony, a consequence of social engineering. The nest contains about 83% worker cells and 17% drone cells. The worker cells contain eggs, larvae, and pupae, one to a cell, which develop into workers. Worker cells are also the primary cells used for storing honey and pollen, though drone cells may also be used. Drones are raised in larger drone cells, as you might have guessed, and are about twice the size of workers (240 mg versus 100 mg, respectively, when they emerge), a result of strong mating competition among the males. Drones raised in worker cells are smaller, the size of workers, and presumably less successful at mating with queens while flying through the air in a scramble competition among sometimes dozens or hundreds or even thousands of other, larger drones.

Drone cells are built on the outer combs and toward the bottoms of combs. This may be a mechanism to control their seasonal production. After surviving winter, the colony begins to renew itself with new bees. The queen increases her egg laying as the foragers discover rich sources of pollen (protein) and nectar (carbohydrate). As new adult bees emerge from their cells and begin contributing to the renewal effort, more and more of the combs are covered with bees and contain eggs, larvae, and pupae. The brood nest expands to the outer combs of the nest and eventually to the drone cells. Conditions for swarming are advancing with the spring buildup, and the queen begins to lay eggs in the drone cells, producing the aerial combatants that will fly out and inseminate the new generation of virgin queens. Producing drones is a serious enterprise and constitutes a major investment. The males contribute little to the colony other than serving as flying bags of sperm, delivering genes to combine with those of queens that will succeed their mother or found a nascent colony. Males have no father. A colony can produce up to 20,000 of them during the spring and summer, each one draining the colony's food resources. Workers, their sisters, cast the remaining males out of the nest in the late summer and early fall and leave them to die. The freeloading males aren't welcome when the reproductive season is over.

The organizational structure of the nest can be thought of as consisting of a hemisphere (Figure 3.4)—except, of course, instead of a solid volume it's restricted to parallel combs with spaces between. The center contains the

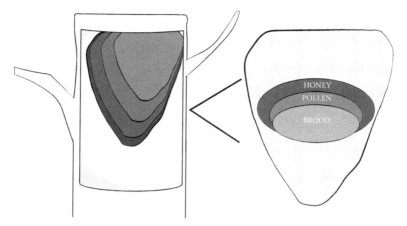

Figure 3.4 Nest architecture. Honey bees prefer to nest inside of dark cavities, such as a hollow tree. A typical colony will contain about eight combs suspended vertically from the top of the cavity. Combs are arranged parallel to each other, like laminae, and separated by about 0.95 centimeters. The oldest and largest combs are toward the center, with the newest and smallest combs to the outside. If you pushed the three combs together, you would create a three-dimensional hemispheric structure. The center contains the brood (larvae and pupae). Pollen forms a thin layer around the top and sides of the brood sphere. Honey forms a cap to the top and sides, filling all available space. If you pulled a comb from near the center of the nest, it would be organized as shown in the figure.

brood, eggs, larvae, and pupae; then a shell of pollen surrounds the brood, and the honey occupies the outer hemisphere, the top and sides of the nest. If you remove a comb from the center of the nest, it will have a lower semi-circle of brood, a thinner semicircular band of pollen around the brood, and honey filling in around the edges. This pattern reflects the coengineering of the nest and social structures. It's a direct reflection of the task division of labor of the workers. Brood is reared in the center of the nest, the most protected part. The temperature is regulated by the workers to 32 to 35°C, optimal for larval growth and development. The queen resides in the middle with the brood, also within the most protected and controlled environment. Young adult workers begin life in the brood area of the nest. When they first emerge, they take up basic housecleaning jobs, cleaning and preparing cells to be nurseries for larvae and pupae. Each cell is carefully prepared before the queen lays an egg. They may also engage in putting wax caps on brood cells containing larvae preparing to pupate. As they age late into their first week of life, they may care for the queen, grooming and feeding her, and begin feeding and caring for the larvae. Each larva is visited by a nurse bee 1,926 to 7,200 times per day and fed during 143 to 1,140 of them (different studies probably had different ratios of nurse bees to larvae in the study colonies). Later, in their second week of life, they may move out of the protective "womb" of the brood nest into more peripheral areas of the nest, where the food is stored. Here, they engage in jobs of receiving and processing nectar into honey, packing pollen into cells surrounding the brood, comb maintenance and construction, and ventilating the nest by fanning their wings, a task performed to regulate temperature and remove moisture as the nectar evaporates and concentrates into honey. Sometime late in the second week or early in the third week of life, they move near the entrance of the nest, where they may engage in guarding behavior or removing dead bodies from the nest (*undertaking*), just prior to leaving the protective environment of the nest to become a forager. Once they initiate foraging, they seldom perform the other within-nest tasks.

Thermal Regulation

Individual honey bees, like all insects, are *poikilothermic*, meaning their body temperature varies with the ambient temperature. They don't maintain a homeostatic temperature like us. We're endothermic and create our body heat internally. Insects use a combination of heating parts of their bodies by increasing metabolism and using the ambient environment (*ectothermy*). That does not mean that they don't control their body temperature. Insects have evolved different strategies and mechanisms to control body temperature within limits for different activities. One of the great engineering achievements of the honey bee has been to gain homeostatic temperature control of the nest; they're social endotherms. They're equipped with social mechanisms for producing, maintaining, and eliminating heat from their nest environment, regulating the nest temperature within a narrow window that ensures the proper development of larvae into adults and the survival of the colony.

Bees don't hibernate but do survive severe winters in the higher latitudes of their geographical range. Some insects seek shelter during the winter where they *diapause*, a condition similar to hibernation. In preparation for diapause and extreme cold temperature, their bodies undergo changes that reduce the formation of ice crystals within the cells, the cause of death from freezing. Some decrease the water content of the body, a kind of freeze-dried solution.

Others produce chemical antifreeze, similar to the antifreeze in your automobile, and wait out the winter months. When I was a graduate student, I did a study of the external anatomy of a bark beetle, *Scolytus multistriatus*, the vector of Dutch elm disease. I wanted to use a scanning electron microscope (SEM) to study the hairs on the external body that serve as touch receptors. I collected some beetles and put them in a −40°C freezer for a day or so before preparing them for the SEM. I retrieved one, thawed it, put it in a total vacuum, plated it with a thin layer of gold, then put it in a vacuum chamber in the SEM, where it was bombarded with electrons. After a few minutes under the electron beam, it started moving. It wasn't dead!

Individual worker bees are thermal wimps compared to the bark beetle. Bees become inactive, barely moving, at temperatures below 7 to 10°C and rapidly perish with temperatures more than 50°C. To fly, they need to warm their bodies to about 35°C, which they do by vibrating their wing muscles to generate heat and by absorbing ambient heat created in the nest. Larvae only develop properly when the brood nest temperature is regulated between 32 and 35°C. Elevated temperatures of only 1 to 2°C can cause serious developmental problems and death.

Colonies of bees in temperate climates face challenges of regulating the nest environment to protect against both excess cold and heat and exhibit social adaptations representing engineering solutions to both. The first step in protecting against winter cold is the selection of a proper nest site: a hollow cavity in a tree that's structurally sound without too many cracks and openings and a south-facing entrance. They seal the inside of the cavity and all cracks and openings, except the entrance, with propolis, making it weather-tight. They may also use propolis to narrow the entrance. The wax combs constructed in the nest and stored food serve to further insulate the nest and provide a substrate upon which bees can cluster to maintain the collective body heat of the colony.

Bees cease to forage and cluster together inside the nest when the ambient temperature decreases to about 18°C. The cluster will form around any brood that may be present and keeps the developing larvae and pupae within the optimal development temperature range of 32 to 36°C. Bees achieve this by clustering on adjacent combs, filling the air space between them. The cluster is less dense in the center where the bees continue to move, feed larvae, consume honey, etc. Their activities and the metabolism of the brood generate heat that's held in the cluster center by the more dense shell of mostly immobile bees surrounding it. The bees in the outer shell typically face head into the cluster. As the ambient temperature decreases, the bees form a denser, smaller cluster; they reduce the surface area over which heat can be lost. The internal temperature is regulated by the diameter and density of the cluster, rather than by generating more heat per individual. However, below about –5°C, the cluster no longer contracts; instead, the bees in the center generate more heat by vibrating their thoracic wing muscles—they shiver and eat more honey. Bees in the outer shell will move into the more central parts of the cluster when their body temperatures drop, where they can reheat their bodies to 33 to 37°C before rejoining the shell layers. They may also feed on honey to restore body energy supplies. In the absence of brood, a normal winter occurrence, the minimum temperature of the inner cluster is kept above about 13°C and that of the outer shell, above 8°C. With long cold periods, clusters may consume all the honey available to them and not be able to move to make contact with honey stored on combs nearby. Honey on combs outside the cluster may be frozen. They starve even when surrounded by plentiful supplies of honey, a not uncommon tragedy for northern beekeepers and their bees.

Using social behavioral mechanisms, a colony is able to survive extreme cold such as occurs in Scandinavia and interior Canada, providing there's sufficient honey available to the cluster. But bees face other thermal challenges as well, especially those that reside in the hot, arid regions of the world. Honey bees can be found thriving from the deserts of North Africa to the tundra of Scandinavia, from the extreme deserts of the southwestern United States to the interior of Canada. How do they survive the heat of the arid deserts?

I spent a career at the University of California at Davis, studying honey bee behavior and genetics. In 2004, I accepted a new position, with new and exciting challenges, at Arizona State University, located in Tempe, a city in the Sonoran Desert abutting Phoenix. I wasn't afraid of heat. I was born and raised in Bakersfield, California, a very hot and arid place, and even Davis gets hot. In fact, the hottest day recorded in the 20 years I lived in Davis was 48°C, the same as the highest temperature recorded in Phoenix during the 15 years I've lived there, so far. The difference is that temperatures higher than 40°C are uncommon in Davis but are daily occurrences during the summer months in Phoenix. Soon after moving to Arizona, I established a bee research lab with around 100 hives of bees and placed them around the bee lab and at other locations nearby. One summer day, I asked one of the technicians to bring to the lab some colonies from one of the out-apiaries away from the bee lab for an experiment we were setting up. It was very hot, and I was naive. He moved the bees the morning of a blistering hot day. When he checked them the next day, they were dead and the combs had melted with honey draining over the bottoms of the hives, a gooey mix of honey, wax, and bees. None of the other hives in the apiary suffered the same fate; they were all just fine, alive and combs intact. What happened?

Adult workers die quickly at about 50°C. Brood dies when the temperature exceeds about 36°C, and wax combs melt at about 62 to 64°C. But combs become soft and weak above about 35°C and will sag and leak under the weight of the honey stored in them. However, bee societies are socially engineered to maintain internal nest temperatures below 36°C even at ambient temperatures as high as 60°C. As the nest temperature increases, some bees begin fanning their wings to generate a flow of air to remove hotter air from the nest. The temperature at which a bee responds and begins fanning is her temperature response threshold. Responses to temperature vary genetically such that within a colony the workers that have different drone fathers (a queen mates with a large number of drones, resulting in the colony consisting of a large number of half-sister sub-families) may respond to different temperatures. Having a genetically diverse collection of workers responding at different temperatures provides a more stable fanning response, leading to better thermal homeostasis. As the temperature rises, increasing numbers of bees respond and begin fanning.

Bees also initiate water collection. They forage for water and place droplets around the nest. The tendency to collect water, like fanning thresholds, varies genetically. The fanning bees line up and produce a circulation of air blowing through the nest and over the water droplets, evaporating the water and eliminating excess heat. Some bees extrude droplets of water from their mouthparts, increasing the evaporative surface area of water. Growing up in Bakersfield in the 1950s, we cooled our house in the same way. We called it an "evaporative cooler" or "swamp cooler." Bees engineered a similar cooling system, and they did it first. When the temperature climbs higher, bees in the nest leave and cluster outside. When they leave, they take the heat from the nest that their bodies have absorbed and eliminate themselves as metabolic heat sources.

When we moved the bees into the bee lab apiary, we failed to provide them with an easily discoverable source of water. Time was critical. They needed to find a source of water quickly to begin the evaporative cooling but failed. We learned our lesson. Be sure you provide a close and easily discovered water source when you move bees in hot weather.

Health

Imagine a hospital nursery containing 7,000 newborns packed together in their bassinets with more than 8,000 nurses visiting each of them more than 1,000 times a day and feeding them more than 100 times daily. Each visit is by a different nurse who recently visited a different newborn—an epidemiologist's nightmare. Diseases would spread rapidly from one to another, aided by the frequent visits of the nurses. Such is the situation in the brood nest of a honey bee colony. Add to that the high density of adults living together in an environment that is managed to be warm and moist, with abundant supplies of sugar and protein, a perfect environment for parasites and pathogenic organisms—and honey bees have many.

Diseases

Parasites and pathogens attack larvae, pupae, and adults of workers, drones, and queens. The worst pathogen of larvae is a bacterium that causes American foul brood disease (AFB; *Paenibacillus larvae*), which kills larvae before they pupate. It melts them down to a thick gooey mass at the bottom of the cell that smells like rotting meat. It can quickly spread through a colony, killing all of the brood and eventually the colony. European foul brood is caused by another bacterium (*Melissococcus plutonius*) and is less virulent. It melts down larvae as well, but they aren't so gooey and smell sour rather than rotten. There are at least two fungal diseases of brood, chalkbrood (*Ascosphaera apis*) and stone brood (*Aspergillus* spp.). Chalkbrood is the most serious and will spread through a colony turning the larvae into flat, hard, chalk-like scales called *mummies* that contain millions of fungal spores. Stone brood is less common but also mummifies larvae and turns them black; they become very hard, stone-like.

Many different kinds of viruses cause a variety of brood and adult diseases. Sacbrood virus turns larvae into a sac full of milky liquid, while the black queen cell virus kills queen larvae and turns them black. It episodically plagues the queen production industry. Deformed wing virus is transmitted to brood by parasitic mites of the genus *Varroa*. *Varroa* transmit diseases by puncturing the cuticle of a bee and feeding on hemolymph (blood) and fat body. When they do this, they pick up virus particles on their mouthparts, then puncture another individual and transfer the particles to the new host.

This is like using a dirty syringe. Deformed wing virus causes the adult bee to have deformed wings and other body parts, shortening their life expectancy; and they're often unable to fly. Other workers also sense they're deformed and reject them from the nest, leading to a reduced worker population and eventual colony death. Adults may also contract viruses that cause bee paralysis, where the bees tremble and are unable to fly. Other bees detect there's something wrong with them and chew on them, leaving many looking shiny and bald, without hairs on their body. It's believed that the chewing bees ingest the hairs, and the hairs scratch and damage the lining of the intestinal tract and open a route for the virus to enter their body. One particularly nasty disease is a microsporidium (*Nosema apis* and *N. ceranae*) that causes the bees to have dysentery. This is particularly difficult in the winter when bees don't get a chance to take their daily cleansing flights—yes, bees have their daily constitutionals. Instead, they defecate inside the nest, spreading the *Nosema* spores and contaminating others, and make a disgusting mess for the bees, and the beekeeper, to clean up.

Mites parasitize honey bees. The tracheal mite (*Acarapis woodi*) is tiny and enters the spiracles of the bee and plugs up the tracheae. Bees breathe through holes located down the sides of their bodies (spiracles). The spiracles connect to a vast network of open tubes, tracheae, and invaginations of the hard exoskeleton, getting ever smaller as they extend deeper into the body. Eventually, the very ends of them, the tracheoles, lie close to tissues needing oxygen and deliver it. Tracheal mites pierce the tracheae with their mouthparts, feed on the bee blood, and reproduce. Eventually, they become so plentiful that they plug up the breathing system and reduce the performance and life span of the bee, like chronic obstructive pulmonary disease for bees. *Varroa* spp., however, are a much more serious problem. Tracheal mites are only an occasional problem, and colonies can usually cope with them. *Varroa* will wipe out entire apiaries if left unchecked. The female mites enter brood cells before capping and lay eggs on the larvae. The mother and the young mites feed on the exterior of the body of the brood, sucking out nutritious bee blood and at the same time inoculating them with pathogenic microbes. They also feed on adult bees by piercing their body wall through the softer tissue lying between the armored, sclerotized segments. The end result is bees with reduced weight and reduced life span. *Varroa* were accidentally introduced to the United States in 1987. Very quickly the populations of feral honey bees living in natural nests were decimated. In California, it was estimated that *Varroa* killed 87% of the feral population within 7 years. It reduced the life expectancy of feral colonies, determined by the continual occupation of a nesting site, from about 3.5 years before the spread of *Varroa* to about 6 months to 1 year. Commercial beekeepers engage in a constant battle with *Varroa* and lose more than 30% of their colonies annually as a direct result of *Varroa destructor*.

Don't cry too hard for the honey bees. They aren't without defenses. Bees have bodies engineered to fight disease pathogens that include an innate immunity, and they have very effective socially engineered defenses.

Individual Immunity
Honey bees have engineered their environment to protect them from disease, predators, and other intruders. Disease defenses are both individual and social. Individual immunity in bees and other insects includes a protective shell around the bee (the exoskeleton), protective coatings within the intestinal tract, and physiological systems that are similar to ours in the blood and cells. The main difference in the physiological system between insects and us is that ours consists of billions of antibodies built by us for specific proteins that invade our bodies, while insects produce very general kinds of proteins that attack broad categories of pathogens, like different bacteria and fungi.

The first line of defense for an individual bee is a hard-shell exoskeleton that's coated with a layer of waxy substances that seals it and protects it from invasion. It's hard on the outside and soft on the inside, the opposite from us. Secretions to the surface have antimicrobial properties as well. As food passes through the gut of a bee, it's contained in an envelope, the peritrophic membrane. The membrane is a semipermeable, complex mixture of chitin, proteins, and polysaccharides (sugars), with a composition similar to that of the exoskeleton. The peritrophic membrane allows nutrients and enzymes to pass through it but protects the gut from pathogenic invaders. Pathogens that pass the barriers confront an innate immune system. Bees don't have a complex adaptive immune system like ours that's highly specific. Each of us has billions of different antibodies (some estimate a trillion), each one able to recognize a specific and different molecule, usually a peptide (short protein), protein, polysaccharide, or lipid. Antibodies are made by our bodies in response to challenges from specific molecules found on invaders, such as a protein on the surface of a bacterium that gives us an ear infection. Our immune system builds and adapts to the challenges it receives. Bees are "born with" broad-acting mechanisms.

The innate immune system of honey bees consists of humoral (in the blood) and cellular (in the cells) responses. The first step in both is the recognition of a pathogen, like a virus or bacterium. Specific proteins called *pattern recognition proteins* (PRPs) are located on the surface of fat body cells and hemocytes (the blood cells of insects) and float in the hemolymph. It's their job to locate and identify invaders. Fat body cells line the abdomen of the bee, and some float in the hemolymph (insect blood). They have functions similar to our liver and are factories for essential proteins, lipids, and other substances essential for bees. The PRPs are looking for specific lipids and carbohydrates on the surfaces of invading microorganisms. These are called *pathogen-associated molecular patterns* (PAMPs). They're like fingerprints for different invading microorganisms. The PAMPs are often distinct between microorganisms belonging to different broad categories of bacteria, fungi, viruses, etc. that may require different mechanisms to eliminate. When the PRPs detect invaders, they induce a response to the pathogen within the blood and the fat body cells by activating chemical signals that result in the production of appropriate chemical defense systems that are involved in killing the invader. The blood cells involved in innate immunity are primarily phagocytes and granulocytes and are similar in function to our own white blood cells. Responses are humoral or cellular. Humoral responses result in the release of antimicrobial chemicals into the blood that cause the breakdown and death of the microorganisms. Some cellular signaling leads to responses such as *phagocytosis*, where a plasmatocyte (a type of blood cell) engulfs the pathogen, takes it into itself, and decomposes it, like some of our white blood cells do. Larger microorganisms, or clusters of bacterial cells, may attract aggregates of plasmatocytes that surround them and kill them. Larger targets such as internal parasites may be encapsulated by hemocytes, forming a multilayer capsule around them. Within the capsule, the hemocytes produce toxic chemicals that kill the parasite.

The physiological systems that attack pathogens in the blood and cells of an insect involve activation of many genes belonging to several gene families. Genome sequence analyses have shown that honey bees have only about one third of the genes involved in individual immunity compared with other, solitary insects. But honey bees have many known pathogens and parasites and live under conditions that attract and nurture them. Why are they so deficient? Social immunity has apparently had a direct effect on the genetically engineered physiological immunity.

Social Immunity

Bees have a kind of public health service within their fortress. The nest structure is engineered as the first line of defense, walled off from the outside
world and lined on the inside with antimicrobial propolis. Guards at the entrance reject non-nestmates, potential vehicles for pathogens, and will also
reject sick nestmates. If you watch the entrance of a hive, you'll often see one
or more bees dragging a bee outside the nest and not letting her return. She
could be sick. Any abnormal behavior triggers their removal. Dead bees are
removed from the nest by the "undertakers." This is a subset of middle-aged
bees, just prior to initiating foraging, that remove the dead bodies. You can
observe them pulling the bodies out the entrance, then taking flight with
them and dropping them many meters away. About 10% of bees die within
the nest; the others die away while foraging. Eliminating them as sources of
disease is good public health policy.

Bees are fastidious cleaners. The first several days after an adult worker
emerges from her cell, she engages in cell-cleaning behavior. Every cell is
cleaned and polished with antimicrobial substances before the queen lays an
egg in it. The food fed to the larvae comes from the hypopharyngeal glands of
the head and contains royal jelly and antimicrobials. Much of the activity of
a bee during the day involves grooming. Bees groom other bees and groom
themselves frequently, removing debris and in some cases parasites from
the surface of the body, though that behavior isn't as well developed in the
western honey bee as it is in the eastern honey bee, *Apis cerana*. Nurse bees
can detect dead and dying larvae and pupae even after their brood cells have
been capped. They uncap the cells and remove the corpses, casting them out
of the nest. This behavior, called *hygienic behavior*, isn't uniformly expressed
in all colonies, posing an important question for bee researchers. Why
wouldn't such an important hygienic behavioral trait be expressed by all colonies? There must be some negative costs to this behavior.

You see many of the mechanisms used by individual innate immunity on a
large scale in social immunity. Nest intruders, such as insect pests, are recognized as non-nestmates by the guard bees; perhaps the nestmate-recognition
cues are analogous to the PRPs and the PAMP is perhaps analogous to a mis-
match of recognition cues found on the body surface and the rejection by
the guard bees. Depending on the type of pest or predator, the colony can
launch different responses. If it's a small insect intruder, the guard could deal
with it herself, sting it, and remove it or simply tug and pull on it to remove
it. If it's a larger predator, a guard could initiate a signal cascade involving

alarm pheromone and substrate vibrations, resulting in guard bees stinging and rejecting the predator. Occasionally, mice or lizards get inside the nest. They're too large for undertaker bees to remove, so the bees will form a ball around the intruder and sting it, then encapsulate it in antimicrobial propolis, sealing it and its potential pathogens off from the rest of the nest. Large wasps present a different problem. They're serious predators of honey bees in some parts of the world and have evolved an exterior armor that bees can't penetrate with their stinger. When detected, the guards send off an alarm and others form a tight ball of bees around the wasps, perhaps like the nodule with the innate immunity cellular response. They compress ever tighter, restricting the wasp from pulsating its abdomen; it can't breathe and dies.

Suites of traits evolve together that include individual and social mechanisms of immunity. Professor Walter Rothenbuhler spent a career at Ohio State University studying the genetics of resistance to AFB. In the 1950s, he obtained some bees from Edward Brown, a beekeeper from Iowa. Brown had been running a wax-rendering plant for decades, taking in thousands of combs and melting the wax out of them. He received a large number of combs from colonies killed by AFB. He stacked these combs outside and let the bees from his apiary remove the residual honey in them before rendering them for wax, thus contaminating his colonies with AFB spores. He'd been selecting for AFB resistance in his bees for decades by basically doing nothing—a natural selection experiment. Colonies from his apiary that contracted AFB died. Those that didn't were used to make new queens that he used to replace queens in his apiary. Professor Rothenbuhler and his students and colleagues studied these bees over three decades, characterizing the resistance mechanisms. They found two social immunity traits and two individual immunity traits that explained at least some of the resistance. The bees were hygienic; they detected dead brood after it had been capped and removed it, a social trait. The royal jelly produced by Brown's nurse bees had strong antimicrobial properties that inhibited the AFB spores from germinating, another social trait. Larvae are only susceptible to disease when they have low body weight, below about 0.7 mg. The resistant larvae grew very fast early in development, spending less time at a susceptible weight. Growth then slowed down so that they were normal by the time the cells were capped—individual immunity. An adult anatomical trait was also affected. Adult Brown bees had a more efficient mechanism to filter out ingested bacterial spores. The honey bee has a three-chambered stomach: crop (foregut), ventriculus (midgut), and rectum (hindgut). Between the crop and midgut lies the proventriculus, a valve. The proventriculus has fine hairs in it that filter particles in the food, such as AFB spores. Brown bees had more. All of these mechanisms have broad-spectrum effects on broad classes of pathogens. None of them were specific responses to specific pathogens like we have.

Selection for Disease Resistance
Let's revisit hygienic behavior. Extreme hygienic behavior can have detrimental effects on colonies, while less extreme expression has been shown repeatedly to be beneficial in protecting colonies from *Varroa* and chalkbrood disease. O. W. Park was a professor at Iowa State University and the doctoral advisor for Rothenbuhler. Park studied and bred for hygienic behavior beginning in the 1930s. He noted that sometimes colonies that demonstrated hygienic behavior got carried away and chewed large sections of comb down to the midrib (the surface that forms the bottoms of the cells on both sides of the comb). Combs would be left with just the capped brood cells drawn out, the rest chewed up and cast out of the nest. My good friend and mentor Professor Harry Laidlaw told me about an inbred line of bees he had selected at the University of California at Davis. Harry said he inadvertently produced a hygienic line he could not keep alive because the bees would chew up all of the combs! Then 30 years later my technician and right hand, Kim Fondrk, uncovered the same trait in some bees he was inbreeding as part of our breeding program at Davis. All colonies derived from queen daughters of one inbred line tore down all their comb, chewing it and "spitting" the wax debris on the ground in front of the hive. I could walk through the apiary and see which they were by the piles of wax. Even lesser expression of this trait could be a high cost to a colony and inhibit the spread of hygienic behavior in some bee populations.

Social Recognition
Social recognition is important in at least two contexts: recognizing nestmates and recognizing the queen. The queen is recognized as a unique individual. Chemical signals produced primarily in head glands of the queen identify her as queen, a royal perfume. She receives special treatment from the workers and is constantly groomed and fed because she produces this signal. But that isn't enough to secure her safety in the center of the fortress. Put the scent on a worker, and she's immediately killed. Put a different queen into the nest, and she'll receive the same fate, even though she produces the pheromone. Are workers recognized as individuals? This kind of recognition

is very unlikely among the thousands of individual workers, but each colony normally has just one queen, making individual recognition possible.

Beekeepers "requeen" their colonies with new queens purchased from commercial queen producers, beekeepers who specialize in raising queens. Some commercial queen producers sell tens of thousands of new queens annually. Although there are many variations of "introduction" methods for new queens, they all begin with removing the old queen. At this point, you can't simply put the new queen in; the workers will fail to recognize her as the queen or as a group member and will immediately kill her. Instead, you place her in a cage where the workers can't come into physical contact with her (but they do feed her through the screen of the cage) and wait some period of time, usually 2 to 3 days, before releasing her to the colony. During the time the queen is caged, she's being "introduced to the colony." When released, she's completely accepted as their own queen, a unique individual within the social group.

But what happens while the new queen is caged? Do the bees learn her individual features? Does she take on some colony essence that identifies her as a member of the group, nestmate cues? She doesn't have contact with the comb while caged. The cages are typically made of plastic with small slits so that the workers can feed queens or of wood with a fine screen, like window screening, covering one side. They also don't directly contact the workers, rubbing against them, picking up nestmate cues from their bodies. At least some of the cues used appear to be genetically determined. A queen's sister is more likely to be accepted directly into a nest than one that's unrelated. But perhaps cues produced by the queen are distributed throughout the nest and learned by workers after the queen is accepted and released. The mechanisms remain a mystery.

Why do bees individually recognize queens? In what context would it be necessary? Queens are found at the entrance of a colony only a few times during their lives, when they take mating flights shortly after emerging as adults and when they leave with half of the workers of the colony and take flight in a swarm during reproduction. In domestic apiaries, queens do, on occasion, make orientation mistakes when taking mating flights and show up at the entrance of the wrong hive, where they're normally rejected. If they gain entrance and begin egg laying, the colony is cuckolded and will expend its energy, raising the offspring of an unrelated (probably) intruder. If and how often this occurs in natural nests are unknown, probably rarely; but the severe genetic consequences of cuckoldry may be enough to maintain it.

Drones have a free pass. They're unlikely robbers, too lazy for that; and honey bees haven't developed a keen nestmate-recognition system for them. They're basically free to immigrate into any nest nearby. In commercial apiaries, they drift regularly between colonies located close by each other. However, they probably seldom drift between natural hives that are usually well separated throughout the environment. They don't collect and transport honey, so they offer no threat other than the small amount of honey they might consume before getting lost again or dying.

Nest Defense

Being contained within an enclosed space, like inside a hollow tree, defines the organizational structure of nest defense. A full-sized colony may contain 20,000 to 40,000 adult bees, mostly workers, and perhaps 20,000 to 30,000 immature bees: eggs, larvae, and pupae. The immature brood is a nutritious source of protein for a hungry and persistent predator and in many cases is more sought after than the sweet reward of sometimes more than 50 kg of honey—just ask any bear or members of some human societies that battle the phalanx of defenders to breech the fortress and acquire protein. But large hairy vertebrates aren't the only worry. They're the occasional threat. Other insects, especially other honey bee colonies, are the constant enemies at the gate, seeking entry to steal the honey treasure or, in the case of wax moths, to lay eggs on the wax combs that will be consumed by their larval offspring. Many species of carnivorous wasps attempt to gain entry and carry away the adults and larvae to feed to their own developing broods.

Being in a contained space limits the entrance areas to the fortress that need to be defended. Bees stand guard at the entrance or entrances to the nest—there are sometimes more than one—and check the credentials of all who enter, determining who belongs and who doesn't. Admission may be based on determining if they're of different species, not a honey bee, or a honey bee nestmate or an intruding honey bee from another colony, a non-nestmate. Admission of non-nestmate bees may result in intruders pilfering honey and returning to their home colony, where they recruit raiding parties to the newly discovered nest. If the nest is sufficiently defended, they fail. If not, the recruited robber bees steal the honey reserves and weaken the colony, laying waste to its hopes of survival. Accurate identification of who belongs and who doesn't is critically important.

A special group of bees guard the entrance. They're middle-aged (13 to 16 days old) and have transitioned from the factory line of brood care at the center of the nest to assume new jobs of fortress security at the nest periphery. Only about 10% of bees ever engage in this task and do it for an average of only 3 to 4 days. As bees or other insects enter the nest, the guards, always alert, move toward them, antennae alerted, and check their credentials. First determination could be whether or not it's a honey bee. If not, it doesn't belong, and aggressive behavior toward it escalates. The guards may bite the intruder and tug at it to remove it from the entrance. Or if it persists in its attempts to gain entrance, the guards will escalate to the highest-level response and sting it, DEFCON 1. If the intruder is a honey bee, the guards may touch it with their antennae to sample the cocktail of waxy-like substances adhering to the body, its nestmate-recognition cues, credentials to gain entry.

Evolutionary biologists recognize two kinds of information that can be used to communicate: signals and cues. Signals are "designed by" natural selection for a specific communication purpose, such as the queen pheromone 9-oxo-2-decenoic acid, also known as 9-o-2 or QMP, for queen mandibular pheromone. Queen pheromone coordinates many social activities. It's important for maintaining the integrity of the colony within the nest and as it flies through the air in a swarm. It serves as an aphrodisiac attractant for drones when queens fly through the air and mate, inhibits the ovaries of workers from developing eggs, and suppresses the construction of queen cells, nurseries for her replacement, thus controlling reproduction. QMP also signals the queen's presence in the nest, eliciting feeding and care from the nurse bees nearby. Over time, the production of the chemical in the head glands of the queen and the ability of workers to perceive and respond appropriately evolved together. Cues are designed for some other function or not designed at all but used to assess some feature, characteristic, property, etc. of an individual. For instance, if you work out in the hot sun all day, your body odor will be a cue to others of your day's activities. However, body odor, a consequence of perspiration and bacteria, was probably not designed to send that message. It's a cue.

Nestmate recognition is based on cues, not signals. The cues come primarily from the wax comb of the nest. Wax is produced by young workers about 5 to 15 days old. They consume nectar and honey (sugar) and convert it into wax scales produced by the wax mirrors, pairs of glands located on the underside of the fourth to seventh abdominal segments. Wax-secreting workers often are inactive, hanging in festoons in open areas of the nest where wax comb is being constructed. Bees have a special spine structure on the hind leg used to remove a wax scale from a wax mirror located between the segments.

The scale is transferred to the mouthparts, the mandibles, where the wax is shaped and placed on the expanding comb surface. The final wax comb is composed of a complex mixture of hydrocarbon compounds including alkanes, alkenes, esters of fatty acids, and long-chained alcohols.

The composition of the comb allows it to be molded but at the same time remain in a solid state at temperatures below about 64°C. It also maximizes strength when used to construct hexagonal cells. However, certain constituent compounds of wax also provide nestmate-recognition cues. The compounds from the wax cling to the hard exoskeleton of the bees. Their bodies are covered with a waxy protective layer of hydrocarbon compounds, similar to paraffin, secreted to the surface. These compounds form a mixture of alkanes, methyl-branched alkanes, and alkenes. The cuticular hydrocarbon blend is of the right consistency to cover the normally porous cuticle and prevent the bees from losing their body fluids and drying out. It also serves as a protective layer against pathogens that may invade through the cuticle.

Neither the wax comb nor the cuticular waxes of bees were designed for nestmate recognition. But elements of both have been co-opted for use. They're blended together on the matrix of the wax combs, mixed with other odors that adhere to and are absorbed by the wax such as floral odors, and then redistributed onto the bodies of nestmates, providing a unique colony bouquet. Individuals learn the bouquet and use it to distinguish between those that are members of their group, nestmates, and those that aren't. The blends of cuticular hydrocarbon cues are heritable and vary from colony to colony, providing unique recognition credentials that are checked by guard bees at the colony entrance.

Guards also respond to movement, color, and odors. Beekeepers move slowly and wear protective clothing made of white, smooth fabric. There's a reason. Guards sitting at the entrance of a hive are very sensitive to movements in their field of vision, especially if those movements are by something that's dark in color. They're also very sensitive to breath, a combination of air movement, odors, and carbon dioxide. Imagine a long history of being robbed by bears tearing into the nest with their powerful claws, sticking their noses in, breathing heavily while they consume the brood and honey. The clothing material is smooth because it gives the bees less to grab with the tiny hooks on their feet while they anchor their stingers into the clothing and skin under it. When I was just beginning to work with honey bees, I entered an apiary wearing dark socks. My pants did not cover them.

I walked by the hives, providing the guards at ground level with a rapidly moving, dark, fuzzy target. Soon my socks were under attack and looked furry from the collection of bees with their feet entangled in the material. My ankles were stung many times, something I regretted even more the next day when they were sore and swollen. The magnitude of the response, however, was a consequence of more than just the visual cues provided. When the responding guards deposited their stingers into my socks and ankles, they left behind the sting, an incredible apparatus for depositing venom and for recruiting additional soldiers. Soldiers aren't guards but are recruited to sting. They're typically older than guards and may forage but aren't dedicated foragers. They're the second line of nest defense.

The sting of the honey bee is an amazing feat of engineering, designed to fully deploy in the absence of the body that delivers it to the target, a fail-safe weapons system (Figure 3.5). Guards and soldiers fly out toward a large potential intruder such as a bear, skunk, human, etc., and may initially butt into it or land on it and bite with the mandibles. If this doesn't discourage the intruder sufficiently, they embed the tip of their stingers under the skin of the intruder and then pull away and fly off, eviscerating the attacker and leaving the sting apparatus to do its job. The stinger of the worker is barbed, unlike the smooth stinger of the queen. The shaft of the sting has three sharply pointed structures that articulate and form a central hollow tube through which the venom is delivered: a central stylet and two lateral lancets. The stylet is fixed in place, and the lancets move forward and backward along the sides of the stylets. This is accomplished by muscles lying at the base of the sting that operate even without the attachment of the rest of the bee. The barbs on the sides of the lancets alternately pull against the skin, progressively ratcheting the stinger deeper into the flesh. Venom passes from the venom sac through the hollow center of the sting and into the wound. The eviscerated worker, intestines hanging, will live several hours. She'll continue to fly around the intruder harassing and confusing it, encouraging it to depart.

It's well known among beekeepers that once you get the first sting more will follow quickly. The defensive response of a colony amplifies soon after the first guards respond, recruiting more and more soldiers. Up to around 10% of a colony of European honey bees (EHBs) may respond to a threat, compared to 50% for Africanized honey bees (AHBs). Defensive behavior tests comparing EHBs and AHBs have shown that AHBs respond much faster and sting test targets many times more. In one test I performed with Ernesto Guzman-Novoa in Mexico, EHBs responded to the test targets in

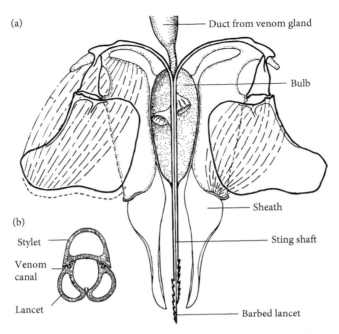

(a)

Duct from venom gland

Bulb

Sheath

(b)

Stylet

Venom canal

Lancet

Sting shaft

Barbed lancet

Figure 3.5 The sting apparatus of a worker honey bee. When in the act of stinging, the worker honey bee inserts the tips, barbed lancets (a), into the skin. The barbs anchor the sting in the skin, then the worker pulls away, leaving behind the entire sting apparatus; flies away; and dies. The stinger continues to function without the bee attached. The paired lancets fit onto the stylet like a tongue and groove and slide up and down along the stylet (b). The lancets are attached to other hard sclerotized parts at the top, which are in turn attached to muscles that articulate the lancets down. Each downward thrust buries the lancet deeper, with the barbs grabbing and holding position while the alternative lancet gets pushed deeper. The alternating downward motion buries the sting deeper under the skin. The bulb is full of venom supplied by the duct from the venom gland. Muscles attached to the bulb pump venom through the venom canal formed by the stylet and the lancets, injecting the venom into the unlucky victim. From *Anatomy and Dissection of the Honeybee* (fig. 12, p. 37), by H. A. Dade, 1977, International Bee Research Association. Reproduced by permission of the International Bee Research Association, ibra.org.uk.

an average of 55 seconds compared to 7 seconds for the AHBs and deposited seven stings compared to 137 during 1 minute. Stinging bees recruit others using a chemical signal called *alarm pheromone* that emanates from glands and secretory tissues at the base of the sting. The pheromone consists of a blend of about 30 chemical compounds, but the primary one is isopentyl acetate (IPA). IPA smells like bananas to our noses but releases a total kamikaze-like defense from bees that sense it. Each additional sting increases the strength of the signal, amplifying the response. EHB responding workers will continue to pursue the intruder up to 50 m, while AHBs may pursue with 10 to 30 times more bees for more than 1 km. EHBs and AHBs also differ in the size of the defensive zone around the nest to which they respond to intruders, with AHBs defending out to more than 100 m and EHBs only a few meters. There are no functional differences in the quantity or composition of EHB and AHB alarm pheromones; AHBs simply are more sensitive to all stimuli associated with defending their fortress, the nest.

Why are AHBs so much more defensive? Africanized bees are a New World phenomenon, created by mixing genes from populations of EHBs and honey bees from Africa. They're the same species, much like crossing a border collie with a pit bull. The stinging responses that all honey bees display in defense of the nest probably evolved as a response to large, vertebrate intruders, like us. In fact, we're probably one of the main causes. Honey and the brood consisting of bee larvae and pupae were, and still are, nutritious sources of carbohydrate and protein. Primitive cultures, and some not so primitive cultures today, robbed bees of these precious resources, leaving the nests broken and the colonies exposed and vulnerable. Africa is the cradle of honey bees and of modern humans, our progenitors. Honey bees and humans evolved together on the savannah of Africa. Humans helped to engineer the niche of the bee, a need for a strong defense; and the bee became an important part of the niche of humans, nutrition.

AHBs Re-engineered the New World

The environmental impacts of honey bees can take place on geological timescales of millennia or timescales of human generations. One of the most dramatic and visible impacts has taken place over my lifetime and is still

happening, the invasion of the genome of Africanized bees into the introduced commercial and feral populations of New World honey bees and the
subsequent spread of hybrid colonies across two continents.

Honey bees aren't native to the New World. Honey bees were first brought
to the United States by colonists who landed in Virginia in 1622. These bees
came from England and were probably of a geographical type adapted for the
climate of northern Europe. Over time, beekeepers preferred bees from the
Mediterranean climate and imported and produced many queens from Italy,
so-called Italian bees, that gradually changed the look and character of North
American honey bees. Feral colonies of honey bees spread throughout the
eastern United States and were referred to as "white man's flies" by the native
Americans—I suppose because they were "stung" by both. They expanded
at least 2,400 km into the heartland of the United States by the mid-1800s,
an expansion of nearly 11 km per year. From there, they were assisted in the
expansion of their range by wagons to Utah and ships to California via the
Straits of Magellan. By 1900, they covered the United States.

EHBs were introduced into Brazil in the mid-19th century. Professor
Warwick Kerr, a well-known and highly regarded Brazilian bee geneticist,
told me the story of how he first learned about honey bees while growing up
in southern Brazil. As a child he was a budding entomologist and fascinated
with bees. He raised colonies of native bees known as "stingless bees" and
harvested their stored honey. Stingless bees are a very diverse group of social
bees found around the tropics of the world and are evolutionary cousins of
the honey bees. They're called "stingless bees" because they lack the stingers
of our honey bees, but that does not mean they're defenseless! Some stingless
bee colonies can attack an intruder with hundreds of workers, each with very
sharp mandibles. They bite and lock onto the skin and may never, ever let
go. You can pull their bodies off, detaching the head, but the jaws remained
locked. They're also known to crawl into the nose of an intruder and bite
down high up the nasal passage. Some species produce a caustic chemical
discharged from glands in the head that causes severe blistering of the skin.
However, young Warwick kept hives of stingless bees that were much less aggressive in their defense.

Warwick grew up in a town in southern Brazil near São Paulo. When he
was about 10 years old, in 1933, he encountered his first honey bee. It stung
him when he collected it. His encounter marked the arrival of the spreading,
feral population of EHBs. This is notable because he lived about 480 km from
Rio de Janeiro, where EHBs were brought to Brazil around 1840. It took them

93 years to spread 480 km, or about 5 km per year, half the rate of expansion in North America. EHBs were adapted to temperate climates, not to the tropical and subtropical climates of southern Brazil; they didn't do well. That was the dilemma and challenge to Kerr.

Honey bees were important for Brazilian agriculture and honey production. But the bees weren't well suited for the environment in which they lived. Colonies didn't build up and survive well and didn't produce much surplus honey. In the early 1950s, as an assistant professor at the University of São Paulo-Piracicaba, Kerr was given the assignment of working to genetically improve the honey bee to be more productive and suitable for Brazilian agriculture. His approach was modeled after the very successful methods of agronomists like Luther Burbank. The approach is to take two varieties of plant that have traits you want to combine into one. Genetically cross them, and you get an F1 hybrid. The hybrids tend to be quite uniform, exhibiting whatever traits have genetic dominance. Next you cross hybrid plants with each other. Their offspring, the F2 generation, will show huge variation, each with a different combination of genes derived from the two original varieties and a mixture of traits. Plants having the desired combination of traits are chosen from the F2 generation to be parents and are crossed with each other. By repeatedly selecting only plants with the desirable combination of traits as parents, you'll soon have the new variety you wanted.

An example might be crossing two varieties of the same species belonging to the genus *Brassica*. *Brassica* is an amazing genus of plant containing many of the foods you find on your dinner table such as mustard, cabbage, turnips, cauliflower, kale, collards, Brussels sprouts, and broccoli. Chinese cabbage and turnips are the same species of *Brassica* but look and taste very different, a result of agricultural selection programs. You eat the leaves of the cabbage and root of the turnip. Perhaps the perfect plant would be one with the leaves of the cabbage and the root of the turnip. Then you could eat the whole plant!

EHBs aren't very defensive and are easy to manage in hives but not very productive in the tropics and subtropics. AHBs are very defensive and more difficult to manage due to their high tendency to produce reproductive swarms where half the bees and the mother queen leave the hive or, when disturbed, will abandon the nest and seek a new one. A beekeeper with 9,000 hives in Sinaloa, Mexico, told me what it was like to use AHBs for commercial beekeeping. Commercial beekeeping in the Sinaloa valley is similar to that of the Central Valley of California—bees on wheels. Hives are loaded on large trucks and transported to agricultural crops to fulfill pollination contracts or to sources of nectar to make honey. Beekeepers depend on their colonies staying in the hives they provide, which is the case for EHBs. However, as this beekeeper explained, after he offloads his hives into the fields, three fourths of them abscond, a huge loss. The bees that aren't in the boxes provided are of no use. However, African bees, unlike European bees, survive well and are productive in tropical and subtropical climates. The goal of Professor Kerr's breeding program was to cross European and African queens and drones and produce a hybrid, then, through additional selective crosses, produce a Brazilian honey bee variety that's gentle, manageable, and productive and survives well in Brazil. Instead, he got the opposite, much like the root of the cabbage and the leaves of the turnip.

Bees were imported into southern Brazil from South Africa and Tanzania in 1956 as part of the breeding program. Hybrids were produced and distributed to beekeepers in the area to test and select their best queens. All seemed to be going well until 1963 when Professor Kerr attended a beekeepers' meeting in southern Brazil. The beekeepers were very angry. Their bees were extremely aggressive, not like the bees they used to have. Professor Kerr told me that he immediately knew what he'd done—and felt terrible. The bees from the breeding program had established a feral population and displaced the weaker colonies of feral European bees that were first there. A subsequent survey showed that by 1963 AHBs had colonized an area of southern Brazil that was the size of Great Britain. (The bees were called "Africanized" because they were derived from the mixing of African genes into the feral European bee population, but genetic studies have shown that the genetic makeup of the population is mostly of African origin.) Africanized bees were on the march, spreading at a phenomenal rate of about 320 km per year, 64 times faster than EHBs. They entered Mexico in 1986, Texas in 1990, Arizona in 1993, and California in 1995. Keep in mind that spreading in this instance isn't just having colonies moving at that rate, but this includes an incredibly high rate of reproduction. As they spread, the number and density of honey bee colonies were increasing, the spread was a consequence of colonies making new colonies that made new colonies, tens of millions of them. The density of feral Africanized colonies has been estimated to be from about seven, in parts of the United States, to more than 100 colonies per square kilometer in parts of Brazil, placing a huge burden on the supply of floral resources shared within the overlapping ecological niches of resident species.

Prior to the spread of Africanized bees, observations of feral EHB colonies were rare in Brazil. They were seldom or never seen at flowers in many areas.

Professor Darrell Posey, an American ethnobiologist, studied the Kayapo Indians in the Amazon basin of Brazil. Kayapo communities have their local beekeeping expert, a position of high rank that's passed from generation to generation. The Kayapo are able to identify scores of different species of native stingless bees and have unique names for each of them. His studies have shown that their system of classification corresponds almost exactly to that used by the best bee scientists: What's a different kind to them is a different species to the scientists. Professor Posey told me that the local native populations mark time since the arrival of the first AHBs into the Amazon region, it was that big of an event! That was in 1972, when the new stinging bees arrived. It changed their lives by changing their niche. They quickly adapted and became robbers of the plentiful AHB colonies inhabiting the rainforest, chopping down giant trees to gain access to honey and brood, altering the rainforest landscape.

High densities of AHBs occur wherever they spread, reconstructing ecological niches. Species of native bees adjust their foraging as a consequence of the intense competition for nectar and pollen from flowers harvested by AHBs. The abundance of native plant species adjusts to the shifts in pollination activity: some becoming more abundant, others less. The floral landscape changes. Populations of feral EHBs, though also invaders of the New World and members of the same species, decline rapidly following the arrival of AHBs in an area; and within a couple of years, their genes disappear with little trace. AHBs restructured the economic niche of beekeeping. Beekeepers complained that their European commercial colonies couldn't make honey in competition with the feral AHBs, resulting in a shift in beekeeping practices beginning with many beekeepers giving up. Beekeeping wasn't fun anymore. Old-guard beekeepers were replaced by a raw, new cohort, beekeepers who had never kept EHBs and didn't know what it was like to keep gentle bees. They adapted to the extreme defensive behavior, altering their management practices. And bees were free. There were so many feral colonies throwing out swarms that all you needed to do was put out a box and soon it would have a colony living in it.

Meanwhile, back in southern Brazil, Professor Kerr initiated a large counter–breeding program to reduce the effects of the feral Africanized bees on the commercial honey bee population. Throughout the 1970s, large numbers of EHB queens were produced and distributed to beekeepers. The idea was that the colonies derived from these queens would produce males, a single colony can produce thousands of them, that would mate with the African or Africanized queens, dilute the gene pool, and reinfiltrate it with EHB genes, making the bees more manageable. Scientists ramped up programs to study the genetics and behavior of the AHBs in an attempt to learn to better manage and control them. The landscape of Brazilian universities changed. New departments were formed to study bee biology and genetics, and new biology departments with bee biology programs spread across Brazil, all a result of the efforts of Professor Kerr. Today, Brazil may have more bee research groups and programs than anywhere else. At a celebration of Professor Kerr's 70th birthday he was presented with a "family tree" of connections of bee researchers to him: his students, his students' students, etc. It contained more than 100 names. The celebration extended over several days in three locations in southern Brazil where he'd made a major impact. Bee scientists from around the world attended. Professor Kerr died in 2018, and I'm sure the family tree continues to grow.

Over time, slowly, the bees in southern Brazil changed, becoming less African in their behavior. Today, the bee industry across Brazil is restored, thriving; the bees are less aggressive, though still not like European bees, and they are productive and well adapted to the climate. A compromise has been reached between the original goals of the breeding program back in 1956 and the African gene invasion that subsequently took place.

As the bees spread northward, the same story played out in country after country. Honey yields declined, people and animals were killed in defensive stinging encounters, beekeepers became discouraged and gave up, and the bee industry was restructured around changes in management practices. Ecological studies showed similar patterns regarding the increased density of honey bees dominating floral resources and niche adjustments by the native pollinators and plants. The Africanized bees were reconstructing the niches of countless species. The feral bees, as they spread, looked and behaved like the African bees with little evidence of crossing with Europeans.

Distinguishing between AHBs and EHBs was the domain of a science called *morphometric analysis*. Techniques had been designed to measure more than 20 individual bee body parts from a colony sample of bees. Data were run through a diagnostic computer program; then, a probability was obtained that the colony was Africanized versus European. The process was costly and time-consuming but was the standard used to monitor the spread of AHBs across Central America, Mexico, and into the United States and was proposed to regulate the transport of commercial honey bees across state lines. The basic concept was that bees that tested morphometrically to be

AHBs had objectionable genes affecting objectionable behavior that we want to exclude, while those that don't test as AHBs are EHBs and don't have bad genes or behavior. If we didn't allow commercial colonies of AHBs to migrate on trucks, we could contain and control their spread. But we can't shut down migratory beekeeping altogether; agriculture depends on it. The California Department of Food and Agriculture (CDFA) planned to stop AHBs on trucks at the border, requiring beekeepers that transport bees across the state line to have their bees tested and certified as AHB-free. My former graduate student, Ernesto Guzmán-Novoa, and I decided to test the efficacy of the method.

Keep in mind that "AHB" stands for Africanized honey bee. It's recognized that they're a hybrid mix between African and European bees, but the degree of hybridization seems to be asymmetrical, biased toward African. As the feral population of AHBs spread, it very rapidly displaced feral EHBs and increased dramatically in colony density. Commercial beekeepers purchase queen bees from queen breeders in parts of the United States that don't have Africanized bees, like northern California, and put those queens into their hives. By doing this, they can keep EHBs. However, colonies can produce reproductive swarms where the old queen, the one purchased from a queen breeder, flies off with half of the workers to establish a new nest and leaves behind a virgin EHB queen that needs to mate with many males to become fully inseminated. Or the queen can die and the nurse bees in the colony will raise a new virgin queen that needs to mate. In either case, the virgin leaves the hive, flies through the air, encounters the drones flying around within about 5 km of the hive, mates with them, and returns. In areas where the feral population is Africanized, the majority of the drones encountered are Africanized, some, likely a minority, may be EHBs from other commercial hives in the area. She stores the sperm of those 20 or so males in an organ called a *spermatheca* to last her egg-laying life of 1 or 2 years. She never mates again. The worker offspring of the new queen reflect the males with which she mated: Most workers are EHB × AHB hybrids (Africanized), while some, perhaps, are EHB, depending on the mix of AHB and EHB sperm she stored.

We artificially inseminated queens using an instrument and techniques developed over decades spanning 1930 to 1970. We inseminated AHB and EHB queens with mixtures of sperm from AHB and EHB drones. We generated 61 colonies that varied in the genetic composition of the workers with respect to the proportion of the genes within the colony that originated from AHB and EHB drones: 0%, 12.5%, 25%, 50%, and 100%. All of these are levels of Africanization that can and do occur in commercial bees. We tested the colonies for stinging behavior and performed the morphometric analyses on 47 of them to test the ability of the method to detect colonies of various degrees of Africanization. We found that the identification method failed to detect all but the most highly Africanized colonies but that colonies that were only 12.5% Africanized were more than twice as defensive as EHBs, making them unacceptable for most commercial beekeeping in California. Colonies that were 50% Africanized had defensive responses indistinguishable from the "pure" AHB sources. Strong defensive behavior is genetically and behaviorally dominant, explaining the difficulty in reducing the defensive behavior of bees in Brazil and the maintenance of strong nest defense in feral bees throughout South and North America.

The research was funded by the CDFA. I contacted Isi Sidiqui at the CDFA, the person in charge of eradicating and regulating invasive species, like AHBs, in California. Sidiqui held an "emergency" meeting that I attended and presented the results of the research. After a brief discussion, he closed his note pad and declared that California would not regulate the transport of bees, marking the end of eradication and control programs in the United States. It was an impossible task.

3.2 Niche Construction

Bees are powerful agents of environmental change, reshaping and constructing the niches of innumerable species that share their habitat— niche engineering. Honey bees engage in niche construction as residential architects, engineers, and builders, constructing nest habitats that wall them off from many challenges of the environment. In the following sections, I'll ask the question whether honey bees engage in the construction of their foraging niche and if they make investments in their foraging environment that pay off in the availability of future resources.

Previously, I proposed that niche construction occurs only in those circumstances where a species benefits from its effects on the environment. It's environmental engineering when the changes to the environment impact other species but not itself. There are two kinds of niche construction: The individuals engaging in it receive a direct benefit, like constructing a nest, or the individuals receive delayed benefits. For example, a colony foraging in a patch of flowers extracts pollen and nectar resources, a direct and immediate benefit that isn't niche construction. But, it may also increase pollination, resulting in more seed being produced and more plants and food resources next year on which it or its relatives may forage, which could be niche construction. To understand this kind of niche construction, it's necessary to understand the concepts of parental investment and reproductive success.

Parental Investment and Reproductive Success

During the course of a lifetime we go through a period of maturation resulting in the achievement of reproductive competence. Once that occurs, we have time and the energy and resources we extract from the environment to invest in reproductive activity such as finding a mate, feeding and caring for offspring, etc. Some of our activities directly affect our reproductive success, while others affect it only indirectly. For example, time and energy spent on our jobs isn't directly reproductive behavior, but it results in money, a resource, that can be used to provide shelter, buy food, purchase health insurance for the family, get orthodontic work for the children, or pay for a college education, all of which may add to the survival and potential reproductive success of our children. Over the course of our lives, we have a total amount of time and resources that can potentially be expended on behavior that's directed toward reproduction; this is our potential parental expenditure (PE). We invest a certain amount of our total potential PE in each of our offspring; this is the parental investment (PI). We assume that higher amounts of PI in each offspring will net a higher reproductive success for them, in terms of survival and their success at rearing offspring (our grandchildren) and hence a higher reproductive success (RS) for us. It follows that the more PE we have over our lives, the greater will be our potential RS. This is the basis of parental expenditure theory derived from economic theory and holds across all animals, including honey bees. However, I must add the caveat that this axiom probably doesn't strictly hold for humans in our modern societies, where RS is no longer the driver of our behavior.

It's obvious that bees change the external environment through their foraging activities and that, in doing so, they alter the ecological niches for other honey bee colonies and for other species. Thus, they're engineering the environment for others. The degree to which they alter the environment to benefit themselves with respect to survival and reproductive success may be considered niche construction as long as it costs them something in the short term, in terms of PE, but results in a net increase in RS. Their behavior has evolutionary consequences for the population. For example, an individual might expend all of her energy and resources finding a mate and providing directly for her offspring—direct investments in reproduction. Or she could expend energy and resources constructing an elaborate nest to the extent that she never gets around to reproducing. It would be stupid if it costs her more to make the nest than she reaps in RS. Such behavior would not evolve or persist in the population as a behavior. Or she could invest in building a nest and get a net benefit through the improved survival of her offspring through improved nest defense, mass rearing of larvae, temperature regulation, hygienic behavior, and food storage, like honey bees. For honey bees, nest construction and the associated behavior of nest living is niche construction.

Nest building and maintenance are obviously niche construction because the bees expend time, energy, and resources to construct and control their immediate environment, protecting themselves from the vagaries of the unpredictable environment outside the nest. But what's the evidence for bees investing in the construction of their external foraging environment? Foraging and collecting pollen and nectar from flowers would be niche construction for honey bees only if it cost the colony in terms of what they could immediately harvest in exchange for what they might be able to harvest later, like next year. For example, patterns of foraging may result in not bringing back the most profitable nectar and pollen in the shortest time or with the highest efficiency. Instead, some bees may visit sparse patches of flowers and "cultivate" them through increasing pollination and seed production. This activity may increase floral resources in the following spring that could be exploited by the colony or by the daughter colony following the loss of the queen through death or swarming. But do they actually have behavioral strategies that evolved because of their effects on altering the environment outside of the nest? Do they cultivate foraging resources for future gain? The answer isn't known, and to my knowledge the question has not been directly asked. Also, the law of parsimony, a central axiom of evolutionary biology, dictates that we interpret evolutionary phenomena by means of the simplest explanation, which in this case would mean looking for explanations where colonies are receiving immediate benefits from the activities of the foragers. But we can look at some peculiarities of foraging behavior and hypothesize that bees may not be making optimal decisions about foraging and recruitment that result in immediate maximum returns of resources to the nest but instead are making investments for the future.

Gifting the Nest

When colonies undergo reproductive swarming, the mother queen leaves the nest with roughly half of the adult bees. She leaves behind a daughter successor along with brood that will emerge over the next days, stores of honey and pollen, the wax comb, and all of the effort expended in constructing the nest, a tremendous dowry for her daughter. She also leaves behind an experienced foraging force with a collective knowledge of the foraging resources within flight range of the nest, a valuable resource. It's obvious that passing the nest from mother to daughter meets the conditions for niche construction. A parental investment was made by the colony in the RS of the next generation.

Resource Fidelity

Bees show high fidelity to single kinds of flowers for periods of single foraging flights or even over many days. Their fidelity is strong even down to foraging only on a specific species or variety of plant. Why? Their fidelity results in more efficient pollination for the plant because the bee brings species- and variety-specific pollen that's capable of pollinating the flowers, but it may not be the most efficient foraging strategy for them. We assume that dedicated foragers learn better how to access the nectaries and anthers of the flower and gain some efficiency in collecting nectar and pollen; they may learn the location and time of day the flowers offer their rewards and save on energy by not visiting empty flowers and learn the visual and odor cues that attract them to the flowers. The general view is that pollination is purely accidental; it just happens because bees seek their own rewards and the flower takes advantage of the bee's greedy behavior. We know that floral species fidelity increases pollination, good for the plants; but we really don't know the degree to which high fidelity may actually reduce the potential return of food to the nest over time when bees pass up profitable opportunities and remain loyal to a floral type or the degree to which this might be part of niche construction.

Foraging on Rare Resources

Professor Norman Gary studied the foraging patterns of colonies of honey bees in agricultural environments. He developed a clever way to track the foraging activities of hives of bees using metal tags and a magnetic retrieval system. Small metal, color-coded, disks were glued to the top of the abdomen of captured bees found foraging in fields containing different crops. Magnets were placed over the entrances of hives located near the fields so that returning foragers were required to pass under the magnets, close enough for the magnets to attract the metal disks. As the bee went under the magnets, her abdomen was pulled up and the disk attached to the magnet. The disks were applied with a type of glue that released the tag once the bee was captured and began wiggling to get free. At the end of the day, the magnets were collected from all of the hives, the tags were removed and sorted by colors representing the fields where the foragers were captured, and a map was made of the foraging patterns of each of the colonies located near the different fields. Gary found that most bees foraged close to home for the most profitable resources; they were maximizing their foraging efforts. They preferred the closest sources of a given plant species. Competition from other colonies limited the distribution of bees, keeping them foraging closer to home. The nature of the competition was probably due to reduced profitability of foraging within the areas of colony forage overlap where another colony had already harvested the flowers. The result is a self-organized, non-antagonistic partitioning of resources among colonies. He also found that some bees in a colony would forage farther for a rarer source of food, passing over closer, more profitable patches of flowers, even if it was less profitable, perhaps a premium for plant diversity. It's assumed that diversity of pollen sources is important for balanced nutrition, but we really don't know to what extent that's true. Foraging bees and nurses apparently can't determine the nutritional value of pollen. Perhaps the foraging behavior is a result of selection for niche construction; the bees may be cultivating a less abundant future resource.

Scout Behavior

If you collect returning foragers at the entrance of a hive and analyze their foraging loads, you find that some bees have collected water or only pollen, others only nectar, many return with loads of both pollen and nectar, but a nontrivial number of bees return empty. When I started looking at the loads of individual bees, I thought that the empty bees were just bees that were unsuccessful at finding resources to collect. Then I did an experiment with Professor Jochen Erber at the Technical University in Berlin, where we measured the responses of bees to solutions of sugar, as explained in chapter 1, section 1.1. Many studies by my students and postdoctoral researchers have applied the same experimental method and have shown that the bees that return empty are the least responsive group when compared with bees that collect only pollen, both pollen and nectar, or only nectar. They're actually programed at birth to be less likely to accept and collect a load when visiting a flower. Our working hypothesis, the most parsimonious, is that these are scout bees searching for the best, most rewarding new resources for the colony to exploit. An alternative explanation is that they're searching for those rare patches of flowers that can be cultivated for future exploitation.

Recruitment Errors

If you watch a forager performing her recruitment dance on the comb, one thing becomes obvious, now that you know how to interpret her dance. Each pass of her waggle dance indicates a slightly different location. There are errors in the performance of the dance transmitting information to the bees being recruited. The recruits visit more than one dance and somehow get a statistical average of the location being communicated, getting them closer to the resource than an individual dance might indicate. But even with averaging, not all bees appear at the location being advertised. They make errors in their interpretation, orientation, and navigation to the source. They aren't perfect. I remember when I was in the army, part of my training was learning to read a map and navigate from one point to another. We would go out on the orienteering course and learn the principles and methods, then take a test. We needed to orient and navigate from point to point using a precise set of locational instructions, the geographical coordinates. We were equipped with a very precise map showing the location and the surrounding landscape and a compass to provide the appropriate direction. We would determine the direction from where we were to the source, the azimuth, and the distance on the provided map. Then we would navigate by using the compass and walking along the azimuth counting our steps—for me, 125 steps was about 100 m on uneven terrain. We could check our position by lining up landmarks from the map. It was far more difficult than it might seem to find a precise point. We often failed. Today, with GPS, it would be a piece of cake. But bees don't have GPS. They most likely don't have a detailed landscape map (my view). They receive from the dancing bee a directional azimuth and distance, neither of which is exceptionally precise.

When the dance language was deciphered, Karl von Frisch emphasized the precision of the information provided by the dancing bee and the precision of the recruits in finding the resource. Von Frisch clearly overemphasized precision; it was unnecessary. The lack of precision became an argument against the validity of the dance language hypothesis. But, as a consequence, there was an apparent need to explain the imprecise nature of the dance and the inevitable errors in locating the point sources of food that were provided in the tests as somehow also being important adaptations. The dance was deliberately imprecise! The most common argument was that it was important for bees to make mistakes and find new resources. That argument isn't accepted today by most. Most accept the fact that it isn't perfect and that by attending multiple dances, each randomly imprecise, the bees get a statistically average azimuth and distance that's better than what's conveyed in any single round of the waggle dance. But, again, one could speculate about programed imprecision as a consequence of niche construction engineering, enabling discovery of new resources that can be cultivated for the future.

3.3 Conditions That Enhance Evolution of Niche Construction Traits

Though there's little empirical evidence for traits of honey bees that can be explained as consequences of evolution for construction of the environment outside the nest, there's some evidence that the conditions exist that support niche construction. If a colony uses its time and energy resources to construct the environment for future reproduction, rather than immediate reproduction, then it must either exclude competitors from using it or have some kind of an advantage for exploiting it relative to competitors. Competitors come in two flavors, those that are of a different species and those that are of the same species. Competition from a different species only affects the evolution of niche construction behavior to the degree the competitors decrease the value of the resource. If they take too much of the constructed resource and leave the builder with too little, then it's wasted effort that only decreases reproduction, and niche construction won't occur. Within-species competition is a struggle of relative value. If you invest in modifying your niche, at a reproductive cost to you, and if your neighbors use your resources and enhance their reproductive success by using them so that the net gain to them is more than yours, then niche construction won't evolve. It pays better to cheat and live off your neighbors' efforts. Why plant your own vegetable garden if you can steal from your neighbors at night? If neighbors can potentially use the resource you invest in, then you must in some way exclude them or reduce their gain—unless they're close relatives.

Neighbors May Be Relatives

Competition from your neighbors can be tolerated if they're close relatives. It's okay if your children live next to you and raid your garden. This can be possible due to the swarming behavior of bees. When a colony swarms, the queen and roughly half of the workers fly away and establish a new nest, usually within a few hundred meters of the nest site she left behind for her daughter. The nomadic colony will easily be within the foraging range of the new one and contains foragers with knowledge of the same foraging resources as those left behind with the new daughter colony. They'll likely compete for common resources, the same ones cultivated by her. The landscape around her may be occupied by relatives from other colonies sharing the common resources cultivated over generations of niche construction.

Monopolizing the Resource

Honey bee colonies can monopolize resources close to the nest and reduce within-species competition. Colony recruitment and foraging are based on the profitability of the resource. Competition for flowers reduces the reward each flower contains and makes foraging less profitable. The greater the distance from the nest to the patch of flowers, the less profitable and the less likely a colony will recruit new foragers to that location or continue to forage there. This is a simple consequence of foraging and recruitment energetics. Bees also learn the location, time of day, floral characteristics, and quality of nectar and pollen for floral patches they exploit. This knowledge of the resource gives a competitive edge to the first colony to find and exploit it, most likely the colony closest to it, again contributing to the nearest colony monopolizing the resource. There's little evidence for active competitive exclusion by bees, but there have been reports of bees from a colony defending a very valuable patch of flowers or artificial feeders by disrupting the foraging of bees from competitive colonies.

3.4 Parting View

I've attempted to present a view of honey bees as environmental engineers. They construct a nest to raise their young, store food reserves, and protect themselves from environmental uncertainties. Their society is engineered to construct and defend the nest and provide for the health and nutritional needs of the colony. Through their foraging activities they change the environment around them, altering the niches of many organisms within flight range. That view is clear. The view becomes much less clear when stretched to include niche construction beyond the walls of their own nest, where the investments made by a colony don't net immediate returns; returns are riskier, delayed in time, perhaps paying dividends only for descendent generations. Perhaps we get tiny glimpses in the form of unexpected behavior such as foraging recruitment errors, scouting behavior, and rare resource preferences; but there's no direct empirical support for this view, so for now the view will remain obscured.

4

The Social Contract

Man is born free; and everywhere he is in chains. . . . How did this come about? I don't know. What can make it legitimate? That question I think I can answer.

Jean-Jacques Rousseau, "The Social Contract" (1762/2016)

We the People of the United States, in Order to form a more perfect Union, establish Justice, insure domestic Tranquility, provide for the common defence, promote the general Welfare, and secure the Blessings of Liberty to ourselves and our Posterity, do ordain and establish this Constitution for the United States of America.

Preamble to the Constitution of the United States,
September 17, 1787

Social insects have fascinated natural historians since Aristotle with the intricate designs of their nests, art forms of nature, the untiring pace of the workers slaving selflessly for the good of the family, and the social order, a hierarchy of royalty and plebeians. The social plebeians are organized into specialized guilds performing the work of building and maintaining the city-nest, providing food for the masses, defending the fortress, and caring for the royal queen. Aristotle recognized this division of labor and wrote about it more than 2,300 years ago. He accurately described a division of labor in honey bee colonies that's based on age. He noticed that the worker bees that foraged had fewer hairs on their bodies than did the bees that worked in the nest. He reasoned that the foragers were the younger bees and that the nest bees were older, a consequence of prepubescence. We now know he had it backward, his human-centric view of bees and behavior had misled him. Foragers are older—the hairs on their bodies break off as they age—but Aristotle's observations led to an understanding of age-based division of labor.

Aristotle viewed honey bees at the entrance of the nest and inferred the social structure from his outside observations. The development of an observation bee hive provided natural historians with a view inside the nest and revealed the intricacies of social organization. François Huber was a Swiss natural historian who was blind from an early age but had an uncanny ability to view honey bee behavior through the eyes of an extremely competent assistant, François Burnens. He used a hive with frames on hinges called a *leaf hive*, where he could work through the combs like thumbing through a book, giving him access to the deepest sanctuaries, revealing the deepest secrets of social life. Through his observations we learned about the reproductive inequities, the division of labor between queen and workers, the bonds of society.

Charles Darwin pondered the social bonds of reproductive inequity in *The Origin of Species*. They provided a serious challenge to him and his new evolutionary theory. How could reproductive division of labor evolve, and why did the workers get into such a cruel contract with the queen, sacrificing their own reproductive freedom to her? And why don't they cast off their chains and raise their own families of offspring? What's the underlying heritable basis of this bad deal, and what kind of selection promoted it? The heritable basis of insect social organization was explained a century later, but the logical and philosophical basis of human social bondage was viewed in depth a century earlier.

Jean-Jacques Rousseau was a political philosopher, one of a string of them who spanned the 17th and 18th centuries, the Age of Reason and Age of Enlightenment, engaged in looking at the moral legitimacy of human societies. Rousseau maintained that there exists a natural state for human beings, one where they are born free and live their own lives, by their own means, and enjoy the natural rights of life, liberty, and property. Men and women become members of a social group voluntarily. Politically legitimate and moral societies are bound together by an unwritten agreement, a social contract, or compact, whereby each member forfeits his or her individual natural rights and thereby leaves a natural state and enters a social state. Individual natural rights are forfeited to the social group in return for protections of those same rights and additional benefits that accrue from cooperation such as increased defense from enemies outside the social group and better opportunities to survive and reproduce.

According to Rousseau, the essence of a social agreement, the foundation of the social contract is as follows: "Each of us puts his person and all his power in common under the supreme direction of the general will, and, in our corporate capacity, we receive each member as an indivisible part of the whole." This act of association forms a public person and can be called a *republic* or *body politic*. The republic is the only moral form of association, giving all individual power and rights over to the whole community. All is equal for all. However, it's not natural to give up personal rights, selfish behavior is innate in us, so other covenants are established to ensure that all members conform. Laws that enforce the general will, the collective will of the people, establish governments to provide services and enforce the contract. Social contracts are tacit; they're implied, not written. The contract is a metaphor for the implicit conditions of harmonious social living, not a written signed document notarized for every citizen. In advanced societies, a written document usually exists that establishes the means by which the general will of the social state is protected and expressed. In the United States, a written constitution establishes how the member citizens vote, a legislative body to pass laws to represent the general will, and a government to execute the laws. It isn't a social contract; it's a law, the "supreme law" of the United States.

In the US Constitution we find the Bill of Rights, the first 10 amendments. They state explicitly the natural rights of citizens that are bound together in a social contract within the state, the United States. These rights are inalienable; they can't be taken away, nor can they be voluntarily given or suspended. By giving up all individual natural rights and giving all your power and resources to the state, you will get back in return those guaranteed rights.

4.1 Origins of Societies

The first human social groups were based on the family. The family is the only natural social state. The social compact between parents and offspring begins at birth and ends at the time the offspring are able to care for themselves; they then enter into a natural state, on their own. However, children may choose to stay home, or close to home, and join a social state with the parents and with each other. To do so, they form a social contract and exist in a social state based on the contract and outside of natural law. These family groups were present in our ancestral species, long before the emergence of *Homo sapiens*, some 200,000 years ago. It's likely that families were at war with each other, taking the life, liberty, and property from each other for their own benefit. It's also likely that families struggled to defend themselves against other natural enemies, such as predatory animals, and to find and kill large prey. As a consequence, families entered into social contracts with each other and with non-family members, forming larger social groups. These events didn't take place just within our own ancestral lineages; many other animal species have similar social structures, for the same reasons, and even though they, like the earliest human progenitors, lack the cognitive awareness to consciously form such bonds, they do so.

But what does all of this have to do with honey bees, the theme of this book? We look at the 20,000 species of bees living in a natural state, visiting flowers, collecting the bounty that their ancestors brought into our world during the Cretaceous, more than 100 million years ago, when there was an explosive evolution of flowering plant species. They build a nest, construct cells for nurseries, work alone, and provide for the young they'll probably never see. Then, we look at the poor worker honey bee, slavishly working away with her worn wings and shriveled ovaries and see her chains. What does she possibly gain? She never lays an egg or sees a job completed; she does only tiny parts of each project, is a tiny cog in the machine of the colony. What ensures that the will of the colony protects the natural rights of the workers? What ensures that they're all born equal with respect to opportunities for life, liberty, and property and treated equally by their social partners? How do they fit into a legitimate and moral social contract?

Of the more than 20,000 species of bees, only a few are socially advanced like the honey bee. When we look at the different kinds of bee societies, we get hints about how they came about and the agreements that were struck among the members. Most bees live completely solitary lives, never interacting with each other, at least not in a positive way, and have no contact with the adult forms of their offspring. They typically construct a nest, one cell at a time either in the soil; the branches of a tree or bush; a dead, rotting log; or solid wood. This is the completely natural state: all against all, only for themselves.

Interactions with other individuals of their species are normally in the context of thievery or usurpation, either defending the nest against another female attempting to take resources or take over the nest or stealing from and usurping the nest of another. The benefit of usurpation is obvious: The nest is already established in a suitable habitat, and there has already been energy invested in burrowing, excavating, and provisioning cells. Some species live in large aggregations of nesting females, each to her own, perhaps occasionally sharing a nest entrance with others. Their compact is, "You take care of your offspring, I'll take care of mine, and we leave each other alone." They may gain an added benefit by incidentally sharing in nest defense because there's just one entrance to defend against thieves, usurpers, predators, and parasites.

Other species of bees share a nest where they jointly protect the brood of all and occasionally care for the brood of another. This is a difficult kind of agreement to enforce the necessary condition of "equal benefits for all" because it's easy to cheat. Some individuals may engage in more care of their own offspring and at the same time get the care investment from their nestmates. They'll have more offspring and make this kind of arrangement unlikely to evolve.

When we try to construct a story of the evolution of honey bee social behavior, it's easy to assume a series of distinct steps in social behavior, each newly derived species more socially advanced than the previous. But social behavior is much messier than that. Many species of bees pass through different levels of social behavior as their colonies develop and mature. For example, species of *Lasioglossum*, one of the sweat bees, tend to aggregate in areas that are good for nesting. Aggregations can consist of more than 1,000 nests. Adult females undergo a winter diapause, something similar to hibernation, often within the nest of their mother, their natal nest. Sometimes more than one female, sisters, reuse the natal nest in the spring. Typically, one of them becomes the primary egg layer, the queen, and the other becomes an auxiliary reproductive, a worker. In time, the queen becomes intolerant of her worker sister and drives her out of the nest; then, the sister will go and found her own nest. The nest is now once again solitary within an aggregation. Within the same species, individual colonies may also begin in the spring by a single female excavating a nest in the ground. She constructs individual chambers for each egg she'll lay and provisions the chamber (cell) with pollen for the developing larva. She closes off the cell and begins the next. At this time, she's in a natural, solitary state. In time, a daughter adult emerges and begins assisting her in nest construction and foraging. The daughter may mate and start her own nest, but most aren't mated. In time, a colony may have a few to a few hundred workers, depending on the species. Like the honey bee, daughters don't normally lay eggs in the presence of their mother, the queen. The founding female remains the primary reproductive, a queen, and the nest is now what we call primitively eusocial. *Eusocial* is a term coined by Suzanne Batra in 1968 for the "truly social" species, where queens and their worker offspring coexist in the nest with a reproductive division of labor between queens and workers and where queens lay more eggs and workers care and provide for the brood. Highly eusocial colonies have a queen that's anatomically distinct, like the honey bee.

In the case of *Lasioglossum*, the final stage of development is primitively eusocial. Individual colonies go through developmental stages that correspond to different states of social organization. However, within the family Halictidae, the insect family to which *Lasioglossum* species belong, different species have different terminal points. Some are totally solitary, and others have different populations of the same species that are terminally solitary, form aggregations, share communal nests, or are primitively eusocial. The patterns of sociality that you see during the development of a single colony or the terminal stages of a single species or the terminal stages of different related species suggest potential evolutionary transitions from solitary to eusocial.

One could imagine that the evolutionary history of eusocial *Lasioglossum* spp. was one where they went from solitary nesters to become aggregate nesters due to competition for adequate nest sites. Then, overwintered sisters reused their natal nests and joined a communal state that led to cooperation in nest construction and brood care. In time, a queen–worker relationship evolved between them, or intolerance evolved where the more dominant foundress ejected the auxiliary helper. An extension of life span for the adults, and perhaps a shortening of developmental time for the larvae, resulted in the emergence of adult daughters during the life span of the founding queen, or queens, some of which stayed in the nest and became workers. Now they have become eusocial. This is just one of many stories that could be told, different hypothetical explanations for the evolution of eusociality, some probably true, others not.

Long periods of evolutionary time make it difficult to find and validate specific evolutionary paths because evolution isn't progressive; it doesn't go deliberately from less complex to more complex or what we think is worse to better. It's fickle, has no memory of where it's been and no goal of where it's going. It only follows the trajectory provided by the environment, with natural selection advancing the better solutions of those that are present at the time. Sometimes the best solution is to revert to less advanced social states, as has apparently happened repeatedly in lineages of bees that went from solitary to social and back to solitary. It depends on the kind of "deal" that's made within the social contract.

When we look at the honey bee, a likely scenario that played out over millions of years of evolution is one that's similar to that of *Lasioglossum*. Ancestral species were solitary nesters. Initially, the females founded a nest, carved out from plant stems or perhaps the trunks of trees, provisioned the cells with pollen, laid an egg, and then sealed the cell. The seasons may have been short, larval development slow, and/or the females short-lived, dying before their offspring completed development to adults. Some species may have immigrated or expanded into climates with longer seasons, development may have accelerated, or females may have lived longer, such that there was sufficient time for the brood to emerge and cohabit in the nest with their mother. Some female offspring may not have dispersed, may have remained behind with their mother and engaged in some degree of caring for their larval sisters. The rest of the story would involve the evolution of larger nests, bigger populations of workers, anatomically differentiated queens and workers, etc., as we see today.

4.2 Social Agreements

Rousseau says that for a social state to persist and to be legitimate and moral, the members of the society must give up their personal natural rights in return for protection of the same rights. They relinquish all of their power and will and accept everyone who joins their society as equal under all of the protections. What natural rights do bees give up? What are power and will to a bee? These are very anthropocentric concepts, based on human experiences, made possible by our large brains and ability to self-reflect and imagine events that might occur in the future, perhaps a unique trait of humans. However, both derive from our evolutionary history that has shaped us, that has been driven by reproductive success. The same is true for the individual and social traits of bees. Power to a human in a natural state may include time, energy expended, ability to defend one's self and family, survival skills, etc., all traits that can affect reproduction. Will is more complicated and has long been pondered by philosophers. What is it, and do we really have free will? In primitive societies, the general will of the group might have been stated as follows: How do we defend our group from large predators and other tribes of humans, protect our families and personal property, divide up the food resources, construct and share shelter, etc.? All of this is directly related to survival and reproduction. But I think in the context of a human society forming a political body, today we think of the will of the people being how much should we pay in taxes; how are our taxes spent on services provided by our government; how should we defend the borders of our country and protect ourselves, family, and property within our society; how should we internally police our own citizens and visitors to insure compliance with our laws, etc.—contemporary social proxies for survival and opportunities to reproduce. One could argue that contemporary humans have psychologically altered their drive for procreation, substituting for it a will for accumulation of wealth to the point that we no longer evolve as a species.

Humans have natural rights by virtue of being living organisms. A bee living in a natural, solitary state has the same natural right to life, as long as a predator or competitor doesn't take it from her; she has the right to be free to reproduce as long as she can find a mate, a suitable nesting site, and food for self-maintenance and provisioning larvae; and she has the right to property, her nest, as long as she can defend it from usurpers. She has the power to find and build a nest; defend herself, her nest, and her developing offspring; and forage for resources. She has but one will, to reproduce. Nothing else matters; all that she does is for that one end. She doesn't look for happiness or things to fulfill her dreams. With a small brain containing only 900,000 neurons, compared to our large brains with 100 billion, she lacks the emotions that we express. Some of us choose not to reproduce because we have other goals to pursue for personal gratification. But, as I argue, my view of the will of humans can, like that of bees, be collapsed to reproduction. The behavior of individual bees and humans within a society has been engineered by natural selection acting on social units, colonies, and family groups, for higher reproductive success; there's no other way, bar divine intervention.

When a bee is in a social contract, she gives up her personal power and will and shares in building and defending a joint nest; defends all of her nestmates and all of their offspring from predators, parasites, and intruders; and forages for everyone in the nest, adults and brood. She also participates in other tasks as necessary to provide for the life, liberty, and protection of property (the nest). Personal reproduction may all be given completely to another nestmate, such as a queen, or reduced and shared with others. The sharing of reproduction is the sticky wicket in social evolution. Darwin pointed it out in the *Origin of Species*. How can reduced reproduction, or sterility, evolve in a population when the whole process of evolution by natural selection is based on favoring those who best survive and reproduce? He saw this as a major difficulty for his fledgling theory. We call this *altruism*. Altruism is when one individual gives up some or all of his or her potential, future reproduction, and gives it to another. This could be a soldier giving his or her life for a country, a requirement of the social contracts of most countries where the power of personal defense is given to the country; a person risking his or her life to save another individual from drowning; or you staying home and helping your mother raise your siblings instead of having a family of your own, as occurs in some human cultures. Each of these apparent "sacrifices" reduces the potential for future reproduction. That's the key to altruism: the cost comes in terms of reproduction potential.

4.3 Evolution of Altruism

Rousseau said he didn't know how the chains of the social state came to bind human populations, but he thought he knew what could make them legitimate. In section 4.1, I provided my view on how the chains of sociality demonstrated within honey bee colonies came to be; here, I show what makes them legitimate in the sense of the social insect contract. The social contract is written in a language with four alphabet characters, G–C–T–A, the building blocks of DNA, and is written in code within the DNA contained within the nuclei of all the cells of the bodies of all bees—and us.

Let's take a deeper look at altruism. Altruism is the fundamental agreement in the social contract of complex social organisms; it defines them. How does altruistic behavior evolve? The calculus of altruistic behavior centers on the concept of inclusive fitness. Fitness is often expressed as the number of offspring produced, but it's actually more complicated than simply counting. *Fitness* is a relative term. There's no such thing as absolute fitness; it's always relative to others, or the fitness of one group versus another. For example, in the human populations of the world, there's a gene mutation that causes red blood cells to take on the shape of a sickle. The mutant form of the gene (s-allele) is most prevalent in parts of the world with a high incidence of malaria such as sub-Saharan Africa. Before we had modern medicine and ways to combat malaria with drugs and insecticides to kill the mosquitoes that transmit it, areas where malaria was common suffered a constant reduction of the population due to the disease. The sickle cell allele reduces the ability of the red blood cells to hold oxygen; shortens the life of a cell, causing anemia; and can cause a disease called sickle cell disease. You inherited two sets of chromosomes, one from each parent. If you get two normal (wild-type) copies of the gene that causes sickle cell disease, one from each parent, you're homozygous (SS) for the normal gene (wild-type allele); you will have normal red blood cells, but you're more likely to die from malaria at an early age. If you get two mutant copies (ss), you're homozygous for the mutation; you will get sickle-cell disease, where your capillaries get clogged with rigid, ill-fitted red blood cells; have many health problems; and die prematurely. If you get one copy of each type (Ss), you're heterozygous and win the lottery against malaria. The organism that causes malaria, *Plasmodium* spp., is a microscopic, single-celled organism that looks worm-like and lives and reproduces inside of red blood cells, destroying them in the process. The red blood cells of a heterozygous (Ss) person provide an environment for the *Plasmodium* that's inhospitable; it can't reproduce, so you're "immune" to malaria, live longer, are healthier, and produce more offspring than others. Your red blood cells are still abnormal but don't clog the capillaries. Your relative individual fitness is higher than that of those with other combinations of alleles if you live in an area with a high incidence of malaria, but you would probably never win a marathon.

There are other levels at which we can explore the concept of fitness. Within populations with a high incidence of malaria, the mutant, sickle-cell allele (s) is relatively more fit than the wild-type allele (S), providing it isn't too abundant. If it's uncommon, few people will have two copies of it (ss). Individuals who have one copy (Ss), will survive better and produce more offspring, and the mutant gene will increase in frequency from one generation to the next; it's the "more fit" allele. If the mutant allele is more common, more people will have two copies of it (ss) and have sickle cell disease, so the wild-type allele will be more fit and result in better survival and more offspring. The population will evolve a frequency-dependent resistance to malaria. A population that has the sickle cell allele in it will be relatively more fit than one that doesn't have it. The percentage of successfully reproducing individuals will be greater, and the population will grow faster, or decline more slowly, than the population that doesn't have it. But this is only the case in the presence of malaria. Without malaria, the sickle cell allele (s) will be selected against due to the reduced survival of the homozygotes, individuals who carry the gene will be less fit due to the negative effects of the gene, and the population will be less fit than populations without the gene.

Group Selection

Selection on groups of individuals was proposed in the early 1960s to account for basically all social behavior observed in nature. For example, male blue jays are noisy and squawk incessantly because they're helping their neighbors assess the size of the population. If there are too many, then some males will voluntarily disperse. Or males are announcing their presence and territory in order to reduce potential conflict from other males accidentally intruding on their space. Individuals in groups that cooperate are more likely to survive and reproduce than those in non-cooperative groups. Group selection is no longer accepted as a general explanation for what appears to be cooperative behavior. The cartoon in Figure 4.1 illustrates the problem with group selection thinking. Lemmings are very small rodents that live in the Arctic tundra; one could easily fit in the palm of your hand. During the winter they live under the snow, feeding on grass and moss, and reproduce; they don't hibernate. Periodically, they overreproduce and increase their population numbers dramatically, so much so that they overeat their food supply. This triggers a migration response in the small rodents, and they migrate in mass. It also leads to a crash in population numbers, to near extinction. It has remained a mystery exactly what causes the population crash. Some believed they must commit mass suicide, for the benefit of the population, though it was never seen to occur. The myth is that they run headlong over cliffs into the water and drown, and they do it for the benefit of the species (group). The myth was perpetuated by Walt Disney productions. Disney made a nature film where the filmmakers faked the mass suicide of lemmings running over a cliff; it never happened but stuck in the minds of many for generations. I saw the film when I was young, and it made a lasting impression on me, as it was intended to do. In the cartoon (Figure 4.1), a mass of migrating lemmings jumps over a cliff, but one is wearing a parachute—a cheater.

Though it may be good for the survival of the population (group) to depopulate and avoid mass extinction from starvation, it isn't good for the

Figure 4.1 Cartoon depicting a lemming migration based on the Walt Disney nature film. In this case the suicidal lemmings are jumping off a cliff, but one has on a parachute, a cheater. It hasn't entered the suicidal contract with the others. Reprinted with permission © CartoonStock Ltd. 2019. All Rights Reserved. www.CartoonStock.com.

reproductive success of an individual lemming. The lemming with the parachute will survive and reproduce, as the others will not. If the trait is heritable, the offspring of the parachute variant will be present in the next generation and will reproduce again. Ultimately, the population will become one where everyone looks out for his or her own interests and suicide will go extinct.

The cartoon is a metaphor for selfish behavior in the face of group selection. It shows why most group selection theories will likely fail. Let's use this argument in something more realistic, the evolution of reproductive and non-reproductive nestmates. Imagine a group of bees that live together in a shared nest. There are many species of non-social bees that aggregate in groups of unrelated individuals living completely separate lives, except for the fact that they share a nest. Assume that environmental conditions are such that more offspring could be produced by the aggregate group if individuals assumed different kinds of roles: Some could lay eggs, others could guard the entrance to the aggregate nest, and still others could forage and bring resources back that are fed to the developing offspring. Individuals who cooperate by assuming non-reproductive roles would leave no progeny, therefore, any heritable variation for the non-reproductive roles would not be represented in the next generation. Only reproducers would be present, no matter what the advantage might be for the aggregate group. This is a fundamental difficulty for the evolution of societies that require differences in reproduction of group members. The genes for these traits reside in the nuclei of cells contained in the bodies of the actors, reproducers or non-reproducers, cheaters or cooperators. Their genes affect their own personal behavior, and they're passed to their offspring. There's no set of genes, no genome, for the group; group behavior depends on the collective genome of the nestmates.

Inclusive Fitness

To understand the concept of inclusive fitness it's necessary to have a "gene's-eye" view. We talk about genes for altruism, but that's just a shortcut for convenience. Altruistic behavior is built from modification of variations in behavior that already exist in the population. Nothing new is needed. Natural selection sorts through the existing variation, changing the frequency of alternative alleles for sets of genes that affect behavior. It's the assembly of sets of alternative genes across the genome that results in changes in what we call altruism. If a mutation exists in a population that makes it more likely that a female will stay home and help her mother, the behavior will evolve in the population only if it increases its relative frequency, more copies of it, in the population from one generation to the next. It can only do that if the individuals who carry it on one of their chromosomes leave relatively more descendent copies of the gene than individuals that have the alternative "selfish" gene. In other words, the sisters that they raise for their mother need to carry the gene or carry it with some non-zero probability. That probability is called *relatedness*. For example, assuming the allele for the altruistic behavior is uncommon in the population, it's likely that the daughter will have just one copy, inherited from either her mother or her father. If she mates, the probability that one of her daughters will carry a copy of her allele, a descendent allele, is one half. That's also the same as the proportion of genes shared in common between her and her female offspring; the same holds for us and our children because they only get half of their genes from us; the rest come from our mates. But she's raising sisters. What's the probability that one of her sisters also has an allele like hers that's descended from the mother? In humans where both our mother and father have two sets of chromosomes, that probability is 50%, the same as the probability that any one of our children shares it. We say that we're related by one half to our children and to our siblings; we share 50% of our genes in common. That isn't the case for bees! Bees, as well as ants and wasps, in the order Hymenoptera, are a special case. Not unique but special (Figure 4.2).

Sex in the Hymenoptera isn't determined by X and Y chromosomes but instead by a single gene that initiates the gender developmental programs. Gender is determined by how many copies of chromosomes the individual has. To become a female, you must have two copies: one inherited from the mother and the other from the father (*diploid*). Males develop from eggs that are unfertilized; they have no father and only one set of chromosomes (haploid)—a sex determination system called *haplodiploidy*. This may seem strange to us large mammals who form monogamous marriages and have a lottery for whether the fertilized egg gets an X or Y chromosome from the sperm of the father, but actually there are many different kinds of systems that determine sex and gender. By convention we call the gender that makes the small motile gametes, like sperm, the male and the one that makes large stationary ones, like eggs, the female. In birds, the female has two different sex chromosomes, Z and W, corresponding to our X and Y. The male has two ZZ. Many populations of lizards, solitary bees, and wasps that are parasites of eggs and pupae of other insects produce no males: females produce females through parthenogenesis. The sex of turtles is determined by the ground temperature of the nest where they develop; they have no sex chromosomes. Some fish change their gender during their lives (e.g., male clownfish can become female), and some organisms, like yeasts, have more than one sex! As a consequence of the haplodiploid system found in bees, sibling females in a nest that share a common mother and father may share 75% of their genes in common, a 75% chance the two will share identical alleles by descent for traits that result in altruistic behavior. We say "by descent" to distinguish from "by chance," which also can occur. You and your brother or sister share alleles for eye color, say blue eyes, because you inherited copies of them from

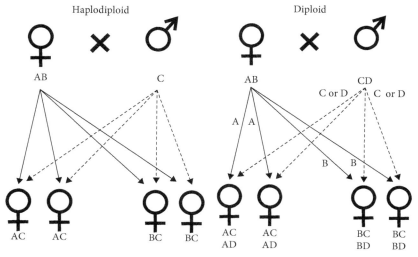

Figure 4.2 Conceptual diagram of genetic relatedness. The figure on the left shows the genes shared in common by descent among sisters derived from the haplodiploid mating shown. The female is diploid: She has two sets of chromosomes and, therefore, two alleles (forms of a gene) at each gene locus, one she inherited from her mother, the other from her father. The male mate has only one set of chromosomes. The daughters from this cross are shown below. Since the female has two sets of chromosomes, each daughter can inherit one of two alleles at each gene locus. In this illustration we're looking at a single gene, and the female has two forms, A and B. The alleles at this locus segregate into eggs randomly, half get the A allele and half the B allele. However, the male has just one allele of this gene, and all of his sperm contribute the C allele to the eggs of the female to make the diploid, female offspring. Only two genotypes are produced at this one locus. If the female laid a large number of eggs, resulting in a large number of offspring, each would share both alleles in common with half of the sisters and only one in common with the other half. Therefore, on average, each shares $(1/2) * (1/2) + (1/2) * (1) = 3/4$ of their genes in common with the sisters. We say they are related by $B = 0.75$. B is a measure of relatedness. The figure on the right shows the case for a diploid mating. In this case, the two alleles at each gene of the diploid male segregate in the sperm just as the two alleles segregate in the eggs of the female. The mating produces four distinct genotypes. In this case each female shares both alleles in common with a quarter of her sisters, a half in common with half of her sisters, and no alleles in common with a quarter. Therefore, on average each shares $(1/4) * (1) + (1/2) * (1/2) + (1/4) * 0 = 1/2$. They are related by $B = 0.50$.

your mother and father, by descent. You and your best friend both have blue eyes because you both inherited the alleles for blue eyes from your parents, and, by chance, they all had the gene.

Individual fitness can be broken down into components (Figure 4.3). For us, we're interested in two components of fitness, classical individual fitness and inclusive fitness. It's through the inclusive fitness effect that altruistic behavior evolves.

Let's return to our example, species of *Lasioglossum*, one of the sweat bees. The species has already evolved to the point where the mother, the nest-founding bee, lives long enough to overlap with her offspring. A daughter emerges and has two choices, disperse and found a nest of her own and reproduce or stay in the nest with her mother and help her mother reproduce. Her choices don't have to be all or nothing; there can be shades of gray in between. The environment is harsh, and nesting sites are scarce; her chances of founding a nest of her own are less than certain. If she stays, she may be able to produce a few offspring of her own, or not, and will increase the reproductive output of her mother. Let's say she, like her mother, has the reproductive potential of producing five offspring of her own. If she stays, she can increase the production of her mother from five to 10 daughters. The deal is sweet for her mother—she can have five more offspring—but what does the daughter gain? Inclusive fitness.

Sex Ratios

The inclusive fitness effect is weighted by genetic relationships. Returning to our sweat bee example, the daughter who just emerged from her cell as an adult will leave more copies of her altruistic genes if she invests her reproductive potential in raising her sisters, rather than her own sons and daughters. For example, if a sweat bee daughter stays in the nest and donates her reproductive potential to her mother, she'll raise sisters that share 75% of her genes in common. If she raises her own offspring, they share just 50% with her. One can say that the relative reproductive value, from a genes-eye view, of a hymenopteran sister is three quarters and that of a daughter is one half. The daughter is likely to leave more copies of any given gene, say one of a set of genes that affects staying and helping at the nest, by not reproducing directly but by rearing sisters. This is a consequence of haplodiploidy. In diploid species like us, the likelihood of passing a gene to the next generation through

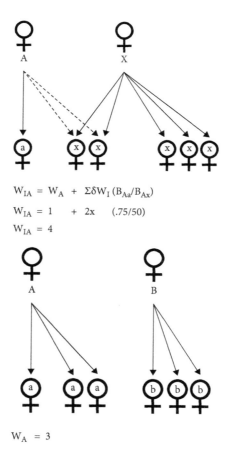

$W_{IA} = W_A + \Sigma\delta W_I (B_{Aa}/B_{Ax})$

$W_{IA} = 1 \quad + \quad 2x \quad (.75/50)$

$W_{IA} = 4$

$W_A = 3$

Figure 4.3 Decomposition of fitness into classical and inclusive components. In the upper figure, I show how we conceptualize the inclusive fitness of individual A who is a daughter of X. They are haplodiploid, and X is mated to just one male, resulting in a genetic relatedness between A and her sisters of $r = 0.75$. I assume that individual A can choose to raise three offspring of her own, or she can stay in the nest and help her mother (X) raise offspring, her sisters. In this case she chooses to raise one offspring of her own (a) and two additional offspring for her mother (x). The equation below the diagram describes the fitness components. The inclusive fitness of A (W_{IA}) is equal to the classical fitness of A (W_A, her own daughter) plus the inclusive fitness component, which is the sum of the changes in inclusive fitness of A associated with each sister she raises. The changes in inclusive fitness are weighted according to the genetic relatedness of individual A to the offspring of her mother, sisters ($B = 0.75$), divided by the relatedness of A to her own daughters ($B = 0.50$). After doing the arithmetic, we find the inclusive fitness of A is equal to 4. The lower figure shows the classical fitness of A and B without helping.

your sister is one half, just like passing a gene through your own offspring. The reproductive values are equal. There's no evolutionary incentive to stay, unless the environment is such that group living itself is favored, such as a shortage of suitable nest sites or a need for more defense due to predator and parasite pressures on the population.

Indeed, there will be more copies of all of her genes in the next generation if she raises sisters. But if she raises the offspring of her mother, she'll also have brothers to deal with. If her mother produces equal numbers of male- and female-destined eggs, unfertilized and fertilized, the daughter will raise equal numbers of sisters with which she shares three quarters of her genes and brothers with which she shares only one quarter. On average, she'll share only 50% of her genes in common with her siblings. This is the same as raising her own offspring. There's no genetic incentive to stay and help her mother. But if she can preferentially raise her sisters over her brothers, then she can once again gain an edge for passing on her genes, and that behavior will increase in frequency; it will evolve. But that requires that she doesn't disperse from the nest and helps her mother and that she can tell the difference between immature male and female eggs or larvae in order to bias her rearing efforts. This whole process of transitioning from being solitary to eusocial is getting more complicated—but not insurmountable.

Not dispersing from the maternal nest is simple. Evolutionary biologists often look at the evolution of tolerance, living together in a nest, to be an initial stage in sociality. However, when you look at insects in general, usually they come from batches of eggs that develop close to each other, and the adults disperse from that spot. Solitary bees and wasps construct nests and lay eggs within them where the eggs develop into adults, and then the adults disperse. Often, one of the offspring remains in the nest, making it hers. Also, often, everyone abandons it. Why? It seems to me that intolerance is more likely the derived trait. What are the conditions that drive the offspring from the nest, seeking to nest all by themselves? Perhaps it's competition and the risk of being exploited by nestmates. But I see little problem in evolving a trait to stay at home in the maternal nest.

Insects and other animals respond to simple stimuli. Birds, for example, give up the food they collect into the gaping mouths of their chicks. They can't help it, it's an instinct. Chicks swapped between nests get cared for. We regularly swap larvae and pupae between colonies of honey bees during population management and routinely in experiments. In some species where females progressively provision their larvae, the larvae produce chemical signals that stimulate their care; they're agnostic about who provides it. Experiments with ants, bees, and wasps regularly swap eggs, larvae, and pupae among nests. So-called slave-making ants raid other colonies of their brood that are then raised in a totally foreign colony, by other slaves. Slave makers take advantage of the instinct to care for brood. I consider joint care of the brood to be an inescapable consequence of cohabitation. Also, when solitary bees are forced to share a nest, a natural division of labor occurs with one becoming the principal egg layer and guard, the other the forager. This has also been shown to occur when queen ants are experimentally forced to build a nest together, queens that normally begin their colony as a solitary foundress. One queen takes on the primary role of excavating the nest. Again, the rudiments of division of labor are inescapable, a consequence of the simple relationships between stimuli and the response of the animal.

Discriminating between male and female larvae also isn't difficult to explain. Many species of bee differentiate between how much food is given to their male and female larvae. Often, the males get less and end up smaller in size. Their size and number are subject to natural selection, to optimize their reproductive value in the population and their mating success. Bees that feed their larvae progressively, like honey bees, can discriminate the sex and age of the larvae, providing for them food of variable quality and quantity. So, I believe all of the necessary preadaptations were there, ready to be shaped by natural selection in developing eusociality.

The queen controls the gender of the eggs she lays, male or female, by opting to fertilize those destined to be female and not fertilize male-destined eggs. However, the workers have a more intimate relationship with the developing brood by caring for them. Once the daughter stays and cares for the larvae, she shouldn't necessarily accept the sex ratio of the eggs provided by the queen but instead skew the investment in males and females. We tend to think about sex ratios as something fixed at 1:1, but that's a consequence of a human-centric view of life. In fact, even human sex ratios aren't fixed at 1:1. The X and Y chromosomes carried in the sperm of males determine the sex of the zygote and the developing embryo. However, it has long been noted that the numerical sex ratio in humans is biased at birth. There are about 105 to 107 males born for every 100 females. However, the numerical sex ratio at the time of reproductive maturity and independence from the parents in most reproductive populations is closer to 1:1. Why? Natural selection on the sex ratio is about populations and is on the investment made in males and females up until they become reproductively capable. This was pointed out by Ronald Fisher in his theory of sex ratio evolution in 1930. The principle is simple.

Within dioecious species (two sexes) reproductive success must be exactly equal between the sexes, a simple consequence of the need for one sperm and one egg at fertilization. However, there could be a huge difference in the variance from individual to individual. A female human over her entire life will release 300 to 400 eggs. A single male produces 200 to 500 million spermatozoa in a single ejaculation, enough sperm in a single ejaculation to fertilize the lifetime egg release of a million females! One male can fertilize all of the eggs of many females, leaving some males without any offspring. The requirement for fertilization requiring intromission (copulation), however, limits mammals to only the number of eggs available at the time of ejaculation, in humans usually one. Some insects, however, can store sperm for years and dole them out to fertilize thousands or millions of eggs.

Each female has the potential of raising a fixed number of offspring; they can be divided between male and female. Those offspring will produce some number of grandchildren. If in a population there are a total of D number of daughters produced, S sons, and G grandchildren produced by them, then each son will on average father G/S number of the grandchildren, and each daughter will produce a share of G/D. Let's say mothers in the population produce twice as many sons as daughters, S = 2D; then the male share of the grandchildren will be G/2D, which is one half of G/D, or half as many as the share of daughters. So, a mother who produces all daughters and no sons will have on average twice as many grandchildren as one that produces all sons. The reproductive value of sons and daughters, in terms of passing on more descendent genes, depends on the sex ratio in the population. All things being equal, you have more descendent grandchildren if the sex of your offspring is biased toward the sex that's in shorter supply. As a consequence, populations will evolve toward an equilibrium sex ratio of 1:1 because at that point the reproductive value of both sexes is equal, no advantage to bias; it's a stable evolutionary equilibrium.

It's actually a bit more complicated than that, of course. The numerical sex ratio will tend to 1:1 only as long as males and females cost the same to produce and the reproductive success of males and females is proportional to the investment in them. In fact, if a male costs twice as much to raise, in terms of time, effort, and resources expended by the parents, then each should return twice as much reproductive success and half as many of them should be produced. It's the investment in males and females that's optimized by natural selection, not the numbers. Which brings us back to the sex ratio of humans at birth. There are about 105 to 107 males per 100 females, worldwide. Sex ratio is optimized for the sex ratio in the population at the time males and females become independent and reproductively competent. Male children have higher infant mortality than females and higher mortality prior to attaining reproductive independence. As a consequence, populations with a 1:1 sex ratio at birth would overall invest more in reproductive females than males, by the time of reproductive maturation. The male bias at birth offsets the asymmetry of investment. A non-elderly adult male in the United States is three to six times more likely to be a victim of homicide than a female and two-and-a-half to three-and-a-half times more likely to die in an accident than a same-aged female. By age 65, there are only about 72 males per 100 females; and, as a consequence, the male bias in the numerical sex ratio declines with increasing age cohorts. The bias at birth offsets some of the higher risk. But how does it get skewed to begin with? Females are more likely to be naturally aborted. One study of 313 miscarriages during the 10th week of gestation showed that 64% of aborted fetuses were female, 36% male, adjustments to the population sex ratio. The ratio of investment of reproductive effort in males and females, sex allocation, is what evolves in populations.

Caveat about Sex Ratios

There have been many twists and turns in the development of sex allocation theory, empirical studies to test it, and our understanding. Sex allocation is broadly seen as a way to empirically test inclusive fitness theory, and most insect sociobiologists agree there's an abundance of evidence supporting it, though some skepticism remains. For instance, as predicted, many populations of the social ants, bees, and wasps demonstrate female-biased allocation sex ratios, though there are additional, competing and complementary explanations for them. Some ant populations where queens mate with multiple males and/or there's more than one queen per colony adjust the within-nest allocation ratios based on the genetic relatedness of the nestmates (more mates and more queens mean lower relatedness and less skew to producing female reproductives). Colonies that reproduce by fission, dividing into pieces, like the honey bee, produce large excesses of males. This is hypothesized to be due to the huge investment each colony makes in each new queen. Workers aren't reproductive; they don't count that way. Reproductives are only queens and drones. Workers are investments, allocations, in queens and drones. In the case of the honey bee, the large entourage of workers that attend the new queen, roughly half the colony, perhaps 15,000 workers, and all of the investment in the nest and brood left behind after the old queen leaves represent an allocation in female reproduction. A huge number of males (up to 20,000) are produced to equal the investment.

Polyandry and Polygyny

Queens in many of the more socially advanced species mate with more than one male, called *polyandry*. And sometimes there's more than one queen, *polygyny*, usually a mother and daughters, or sisters, resulting in a reduction of the relatedness of workers to the female offspring of the queen. In the case of the honey bee, the queen mates with a large number of males. Estimates vary, but some are in excess of 20. As a consequence, most workers are half-sisters sharing just 25% of their genes, not supersisters that share 75% (Table 4.1).

Table 4.1 Table of Relatedness and Terminology Comparing Diploid Species Like Us with Haplodiploids Like Bees

Pedigree	Diploid terminology	Haplodiploid terminology	Diploid relatedness	Haplodiploid relatedness
Same mother and father	Full sister	Supersister	0.50	0.75
Same father different mother	Paternal half-sister	Cannot occur naturally	0.25	—
Same mother different unrelated fathers	Maternal half-sisters	Maternal half-sisters	0.25	0.25
Same mother different fathers that are brothers	Maternal half-sisters	Full sisters	0.50	0.50

Note. Column to the left gives the pedigree relationships of individuals. Center two columns provide the terminology used to describe the relationship for diploid species and haplodiploid species. Column on the right shows the genetic relationships (proportion of genes shared in common) between individuals.

We use the term *supersister* to recognize that all sisters of the same father inherit the same set of genes to distinguish them from full sisters, like our sibs, who share a father but only share half of their genes in common. As a result, average relatedness of honey bee workers to sisters approaches 25%, the same as their relationships with their brothers, leaving no incentive to stay and skew the sex ratio (Table 4.1). How did this come about? How can we justify it with inclusive fitness theory?

As discussed, it's most likely that honey bee progenitors were singly mated reproductive females that occupied their own nests, evolved extended life spans, and developed intimate relationships with developing larvae through progressive feeding. The next phase probably involved natural selection operating on the inclusive fitness component of individual fitness, resulting in some females forming a social compact with their mother and remaining in the nest and rearing brothers and sisters. Over time, the workers and queens became progressively more specialized and evolved anatomical differentiation between them, the queen becoming more and more an egg-laying specialist and workers adapted less for reproduction and more for performing functions of defense, nest construction, food collection and storage, brood care, etc. As the workers became more specialized on worker functions, they had fewer and fewer options to return to a solitary state and/or start their own colony if the deal changed for them. At some point they reached what Bert Hölldobler has called the "point of no return" in social development. At this time, selection on whole colonies dominates the fitness of the individual workers; all that remains is the inclusive fitness effect. As hypothesized by Charles Darwin, colonies that had heritable organizational structures that were better with respect to surviving and reproducing than alternative structures passed more genes to the next generation, and those favored structures evolved in the population. The genes for those social structures, behavioral, anatomical, and physiological, reside in workers. The genes were sequestered in the ovaries of queens and the sperm of their many mates, the repositories of genomes that when expressed in the workers produced a society with a division of labor. Colonies that had more genetic diversity, a consequence of queens mating more times, probably were better able to resist diseases, had a more efficient division of labor among workers, had higher survival, and produced more reproductive males and females, so multiple mating evolved. Workers continued to benefit due to the increased reproductive success of the colony, but their genetic interests now were approaching full agreement with the queen's.

4.4 Social Benefits

Natural selection forms and enforces the social contract through the inclusive fitness component of individual fitness. Colony-level selection additionally establishes the conventions of society that determine the social structure, a social compact. Workers assume a set of tasks, jobs that they perform for the functions of society. These benefits are similar to the guarantees provided in the Preamble of the Constitution of the United States shown at the beginning of this chapter:

1. **Establish justice and insure domestic tranquility**. Honey bee societies guarantee equal treatment of all members. Workers have no individual identity; they either belong to the colony or are rejected. It's necessary that individuals fulfill their social responsibilities by yielding all of their power and will to the state (colony); otherwise, it will collapse into anarchy. Worker honey bees retain the ability to lay eggs that will become males. This is a natural state in the absence of the queen. Colonies in that state collapse, every individual only working for herself and pursuing her individual will, reproducing. In the presence of the queen some attempt to cheat, taking the services provided by the colony but working for their own personal reproduction. Nurse bees police the egg laying of other workers. They inspect cells containing eggs and are able to tell the difference between eggs laid by the queen and those of workers. They remove and consume worker-laid eggs. Egg-laying workers are likewise detected, harassed, and often expelled from the nest—no social cheats allowed. Domestic tranquility is further maintained by the pheromones of the queen. Queen mandibular pheromone (QMP) suppresses anarchic behavior by suppressing ovary development and egg-laying behavior. QMP also has a "calming" effect on the colony. Colonies without queens have a noticeable increase in activity, fanning behavior with workers running on the comb. Eventually, there's a complete breakdown of social structure, discussed in Chapter 6.

2. **Provide for the common defense**. Society provides for a defense of the nest (Chapter 3). Honey bee colonies have a group of bees that guard the entrance, denying entrance to bees that aren't part of their colony. Honey bees have evolved a nestmate-recognition system whereby they check the chemical credentials of bees at the entrance, their passports,

rejecting entrance to non-nestmates. If the threat of an intruder escalates, increasing numbers of bees are recruited, drafted, into the defense corresponding to the level of the threat.

3. **Promote the general welfare**. Colonies provide *public welfare* services that benefit all of the inhabitants, such as housing, by selecting, constructing, and maintaining a nest that protects the colony from the external environment, thermoregulation, food distribution by providing for nutritional needs through the supply of stored honey and pollen, and childcare by caring for the larvae in a common nursery (Chapter 3). Colonies also have *public health* services. Undertaker bees remove the bodies of dead adults. Unhealthy bees are expelled from the nest, no longer socially acceptable. Hygienic bees uncap and remove dead larvae and pupae, removing the threat of the spread of diseases. *Public works* services are also provided. Many bees that are in about their third week of life clean the nest, waterproof the walls of the nest with plant resins, and construct and rebuild comb. Bees also forage for water and bring it to the nest. Water is used for diluting honey to feed to larvae and for thermoregulation. The queen is a public servant of the colony. She was domesticated by the workers over millions of years and became an egg-laying machine, working day and night, depositing eggs into individual cells, raw material for shaping a society.

4. **Secure the blessings of liberty to ourselves and our posterity**. Honey bee societies provide for an orderly transition of resources from one generation to the next. The act of swarming not only divides up the adult population of the nest but also provides for the next generation to have the resources and a workforce already specialized to provide the social services necessary to succeed. The old queen leaves with about half of the population to establish a new nest and leaves behind for her daughter the nest with all of its resources and, importantly, all of the workers who are already engaged in behavioral tasks that support the society.

4.5 Parting View

Societies have fundamental structural needs to succeed; otherwise, they fail. This is true for insects and humans because we have the same natural will that must be controlled, selfish interests associated with reproduction.

Behavior of individuals within societies has evolved by a process of reverse engineering. Colony-level natural selection acts on existing social structures through the inclusive fitness of workers that are consequences of the activities and interactions of all of the members of the society. As a consequence, characteristics of the individual members (anatomy, physiology, behavior, etc.) change to fit the needs of society; a social contract is formed that's written in the DNA of each worker: the social contract of the superorganism.

5

The Superorganism

Victorians abound in such phrases as the "struggle for existence," "survival of the fittest," "nature red in tooth and claw," and disquisitions on the unrelenting competition in the development, growth and behavior of all animals and plants. This struggle, as you know, was supposed to constitute the very basis for the survival of favored forms through natural selection . . . but we would insist that it depicts not more than half of the whole truth. To us it is clear that an equally pervasive and fundamental innate peculiarity is their tendency to cooperation.
William Morton Wheeler, *Social Life among the Insects* (1923, p. 3)

In 1911, William Morton Wheeler proposed that ant colonies should be considered organisms. The organism as a metaphor for a society wasn't new, having been used since at least writings of the Middle Ages, such as in the term *body politic*. The body politic comprised all the people in a particular country as the body with the sovereign leader as the "head" of state. Thomas Hobbes in 1651 illustrated his metaphorical *Leviathan* as the frontispiece of his book, a body composed of all the people of the land, arm raised high brandishing a sword that enforces order, with a monarch's head (Figure 5.1). Honey bee societies often were referred to as "cities" or as "domains" ruled by a monarch, a king for Aristotle in the fourth century B.C.E. and a queen for Charles Butler

Figure 5.1 Hobbes' Leviathan. In 1651, Thomas Hobbes published *Leviathan, or the Matter, Forme and Power of a Common-Wealth Ecclesiasticall and Civill.* Regarded as one of the earliest and most influential treatments of social contract theory, the etching shown here was created by Abraham Bosse as the frontispiece for the book. It depicts a body politic, the body of a powerful monarch composed of the bodies of the members the political community. From Leviathan, by T. Hobbes, 2002, Project Gutenberg. Retrieved August 16, 2019 (http://www.gutenberg.org/ebooks/3207). In the public domain.

in *The Feminine Monarchy*, the first English-language book on beekeeping, published in 1602. In 1860, Herbert Spencer, a well-respected English philosopher, biologist, and sociologist, perhaps best known in biology for coining the phrase *survival of the fittest*, likened society to an organism and presented a catalog of analogies mapping functional units of human societies onto the anatomical framework of organisms—for example, likening the office of Parliament to the brain of an animal, though one would certainly question the analogy today. Hanging analogies on the framework of metaphors has long been practiced with mixed results. Sometimes it leads to interesting questions or perhaps a new view that wouldn't have been realized without drawing the parallel comparisons. Likening the planet earth to a living organism builds awareness of our need to protect it and respect for the intrinsic global processes that can and have been disrupted by our own activities. It has also led to studies of how global heat is transferred and how ocean currents and rain forests affect weather cycles, carbon dioxide is sequestered and released, etc. But, as my dear friend and philosopher Sandy Mitchell once told me, a metaphor is just a metaphor. There are limits to their usefulness.

But Wheeler didn't see the ant society as a metaphor; he saw it as a true entity with the characteristics of an individual organism. In 1911, he defined an organism as

> a complete, definitely coordinated and therefore individualized system of activities, which are primarily directed to obtaining and assimilating substances from an environment, to producing other similar systems, known as offspring, and to protecting the system itself and usually also its offspring from disturbances emanating from the environment. The three fundamental activities enumerated in this definition, namely nutrition, reproduction, and protection, seen to have their inception in what we know, from exclusively subjective experience, as feelings of hunger, affection, and fear, respectively.

He continues with

> The most general organismal character of the ant-colony is its individuality. . . . it behaves as a unitary whole, maintaining its identity in space, resisting dissolution and, as a general rule, any fission with other colonies of

the same or alien species. This resistance is manifested in the fierce defensive and offensive cöoperation of the colonial personnel. (1911, pp. 308–310)

He first used the term *superorganism* in 1928. The term stuck and has been a part of the vocabulary of social insect biologists ever since.

The epigraph at the beginning of this chapter shows that Wheeler set the social insects apart because of his dissatisfaction with Darwin's theory of evolution by natural selection. He claimed that there must be more to the existence of insect societies than accounted for by Darwin. Metaphors for natural selection such as "nature red in tooth and claw" and "the survival of the fittest" painted a picture of nature in constant conflict for survival and reproduction, individual pitted against individual. How can you explain the evolution of a cooperative society? The superorganism wasn't subject to the same laws of natural selection; he needed insect societies to somehow be different, to lie outside of the action of Darwinian selection.

5.1 Darwin's Theory of the Superorganism

Wheeler wasn't the first to cast doubt on the ability of Darwin's natural selection to build cooperative societies. Darwin's own skepticism preceded Wheeler. Darwin in 1859 felt that social insects presented different kinds of difficulties for his fledgling theory. The first was in explaining the evolution of instincts that result in the construction of honey bee comb that's "absolutely perfect in economizing labour and wax." This is "effected by a crowd of bees working in a dark hive" (Darwin, 1998, pp. 339, 348–349). He recognized that there were no task masters orchestrating the coordinated activities of comb building. He struggled first with the evolution of perfection, considering such questions as "How do you get to perfect comb, or highly complex functioning organs like the human eye, when natural selection grinds slowly through naturally-occurring, heritable variation?" The intermediate steps from no comb to perfect comb or no eyes to highly developed eyes must have had some kind of advantage along the evolutionary course, though perhaps difficult to imagine. In the case of comb construction, there's the additional problem that comb construction depends on instinctive behavior contained within the worker bees, who don't reproduce. Instincts are inherited. The workers demonstrate the instinct but don't reproduce, while the queens and drones reproduce but don't demonstrate the behavior. His theory requires differential survival and success in reproduction. How can you evolve traits in worker castes that don't reproduce? He pointed out the additional difficulty of explaining the evolution of multiple castes such as those that occur in ants. The reproductive queens are anatomically and behaviorally different from the workers, as is the case for the honey bee. But many ant species have the worker caste further divided into different anatomical castes that differ even more from each other than queens do from workers, such as tiny minims, huge scary-looking soldiers. How can they evolve?

Wheeler rejected Darwinian natural selection and offered no alternative. He just left it open as a mystery. Darwin, however, did offer an explanation, a theory of the superorganism, though he didn't call it that. He said that selection acting at a higher level than the individual, the family group, could explain the evolution of sterility and caste differences. Families that are composed of parents and offspring that have certain sets of characteristics might survive better and produce more successful reproductive offspring than other families composed of offspring with different features. If heritable, the favored family composition would increase in the next generation, while the less favored family type would decrease. The population would evolve. Animal breeders regularly take advantage of this. Breeders of milk cows measure the milk production of the daughters of a breeder cow. Daughters of those that have the highest average yields of their family are selected to breed in the next generation; other families go to the slaughterhouse. Darwin didn't elaborate on just how this could occur—that needed an understanding of genetics, inclusive fitness, and an additional 150+ years to be resolved—but he was correct in the broad sense.

Darwin also offered an explanation for how a crowd of bees working in the dark could perform apparently coordinated behavior, comb building, without a central set of controls. This is another important part of understanding how superorganisms came to be and evolve. Honey bee colonies may contain 40,000 bees, acting together in what looks like coordinated behavior. But there's no blueprint, no working group foreman (actually forewoman) overseeing the project or central control system. Again, Darwin was more than 150 years ahead of his time in pointing out the difficulty with complex social behavior emerging from the crowd of individuals. He conducted experiments and explained how a perfect comb could be constructed from basic instincts of workers that behaved based only on the information they were able to gather through their own comb-building activities. They don't need centralized control. I'll deal with more of the specifics in the sections that follow.

5.2 Wheeler's Superorganism

Wheeler first used the term *superorganism* in 1928 in connection with analogies of the functions of insect societies and organisms, offering nothing beyond a conceptual way of thinking. As an ant biologist, his analogies were ant-centric, of course. You can also see parallel analogies with the honey bee and other social insects. The most general characteristic of an ant colony is its individuality, behaving as a unitary whole. It fits his definition of an individual organism. Ant colonies are also idiosyncratic, like us. Each is composed differently and behaves differently. This is something that Jennifer Fewell and I noted about colonies of honey bees. We did a series of experiments in Davis, California, where we looked at the foraging behavior of honey bees from two strains selected for their pollen-foraging and -storing behavior. We wanted to control everything about their genetic makeup and their hive and foraging environments as carefully as possible, and we wanted to replicate the environments. We set up four screened tent cages to contain the bees; each cage received one colony. The colonies were established in small hives as identically as we could possibly make them with respect to numbers of bees and the composition of the combs we provided. Each colony was composed of the same mixture of bees from the two genetic strains, and all of the bees within a strain were sisters. So, we minimized the genetic differences between colonies in the different cages. All of the cages faced the same direction and received identical exposure to the sun throughout the day. Foragers flew out of the hives and collected pollen and nectar from identical feeding stations placed in identical positions within the identical cages. We recorded when the bees foraged and what they collected. The four colonies were completely different. Each had its own daily "wake up" time when foraging was initiated, and they foraged with different tempos. They were idiosyncratic.

Biogenetic Law

The train from Berlin to Munich, Germany, passes through the town of Jena. The first time I made this trip was 1996, six and a half years after the Berlin

Wall came down and East and West Germany were reunited. East Germany still bore the scars of 50 years of neglect under the East German government. The contrast between West and East was staggering. I remember passing slowly through town, looking up, and seeing an old run-down building of the Friedrich Schilling University. Across the facade over the front entry to the building was painted an evolutionary tree of life, like a large bush. To the left of the tree was written "Ontogeny" to the right "Phylogeny." I remember a sense of awe came over me when I first saw it, bringing thoughts of how Jena was once an intellectual center, especially at the turn of the 19th century with Friedrich von Schiller (poet, philosopher, playwright, physician, historian), Wolfgang von Goethe (poet, biologist, playwright, novelist), and the philosopher Friedrich Hegel. The university was founded in 1558, making it one of the 10 oldest in Germany, and was renamed the Friedrich Schiller University in honor of one of its most famous faculty. It was frequently visited by one of my heroes of biology, Alexander von Humboldt, a close friend of Goethe. But the inscription recognizes the work of another famous member of the faculty from the middle to late 19th century, Ernst Haeckel.

The inscription encapsulates what became known as the *biogenetic law*, "ontogeny recapitulates phylogeny." Haeckel proposed that while embryos develop, they pass through the different stages of the animal's evolutionary history. Embryos of amphibians, birds, and mammals, including us, look remarkably similar during the earlier stages of development and do not diverge in appearance until the later stages. So, Haeckel believed that as we develop in the womb of our mother we pass through our evolutionary stages; we become a fish, a reptile, a rodent, a chimpanzee, and then at the end a human baby. Additional evidence for this in humans supposedly came from the occurrence of a "tail" during embryonic development that's resorbed by the fetus, usually. Human fetuses also go through a stage of development where they're covered with hair, lanugo, that's usually shed before birth but not always. And the prehensile grip of an infant's hand is incredibly strong for its size, and the feet assume a natural prehensile grip position, good for living in trees. These characteristics were explained as atavisms, leftovers, from our evolutionary history played out in development, recapitulation. This theory was subsequently shown to be wrong and replaced by Baer's law, but it was broadly accepted for some time.

Karl Ernst von Baer preceded Haeckel by about half a century, working in the early to mid-1800s. Baer noted that similarities across species during embryonic development only reflect the evolution of embryos, not a reca-pitulated evolutionary history of adult forms. The reason vertebrates are so similar in the earliest stages and less alike later is that it's hard to change early developmental processes because so many developmental steps come after them; small changes early result in huge differences later or are fatal. It's easier to change the later stages of development to end up with, say, a chimp or me. So, ontogeny does reflect the phylogeny of developmental evolution, but it doesn't recapitulate it.

The biogenetic law of Haeckel was accepted by Wheeler, and he invoked it as evidence that ant colonies were organisms. Like us, as they develop they recapitulate their evolutionary history from solitary ancestors to the com-plex social organisms they are today. Most queen ants initiate a nest alone. They lay some eggs, just like so many less advanced solitary insects. When the first workers hatch, the colony is very small. Many primitive species of ants never form large colonies and always stay small. However, the more ad-vanced ants like the leaf-cutting ants form huge colonies consisting of many different forms of workers. As colonies of the more advanced species get larger, the relative percentages of the different types change. For example, when they're small they only produce very small worker ants, but as they grow they produce larger workers and more and more soldier ants to de-fend the nest or specialized ants for cutting leaves, storing honey in their bodies, etc.

Being "wrong" in science is inevitable. In his autobiography, Darwin talked about the likelihood that his theory of pangenesis, a rather strange theory of inheritance before the genetical basis of inheritance was discov-ered, was wrong.

An unverified hypothesis is of little or no value; but if anyone should be led to make observations by which some such hypothesis could be es-tablished, I shall have done good service, as an astonishing number of isolated facts can be thus connected together and rendered intelligible. (p. 38)

I cannot remember a single first formed hypothesis which had not after a time to be given up or greatly modified. (p. 44)

Such is the nature of science as a self-correcting process of organized thought, a way of knowing.

Germ Plasm Theory

Another highly esteemed 19th-century German biologist had a more en-during theory of developmental biology. August Weismann was an evolu-tionary biologist working in the Albert Ludwig University of Freiburg in the mid-19th to early 20th centuries. He's best known for his germ plasm theory. He proposed that multicellular organisms are composed of two ge-neral kinds of cells, those that form the germ plasm (make eggs and sperm in animals) and those that form the soma (other tissues like heart, liver, skin). Once a sperm fuses with an egg, they combine their nuclei and form a zygote. This is still one cell. The zygote then starts dividing, doubling the number of cells with each division. Early on, different groups of cells become destined to form different tissues in the developing embryo. The germ cells that ulti-mately make eggs and sperm are set aside very early, sequestered from the other body cells, the somatic cells, and undergo differentiation into sexual tissues later, like the ovaries and testes. This protects the germ cells from any influence of the other body cells. Only the germ cells are able to pass on her-itable information to offspring through eggs and sperm, all the modifications of other tissues, such as tumors, lacerations, or tattoos, are blocked from passing to the offspring. This theory was in direct opposition to Lamarckian theories of the inheritance of acquired characters—for example, cut the tail off a rat and its offspring will have short tails—that were still prevalent at that time and became one of the most important concepts in evolutionary devel-opmental biology.

William Morton Wheeler believed that ant societies were organisms be-cause the reproductive (germ) and non-reproductive (soma) parts (tissues) are separate like other multicellular organisms. Queen ovaries and the sperm from their mates that they contain in their sperm-storage organ constitute the sequestered germ line of the colony, while the sterile workers are the soma, the cells forming the body of the superorganism. This relationship in ants was also noted by August Weismann nearly 20 years earlier. Like our body cells differentiate into specific kinds of tissues with different functions, workers become differentiated into distinct castes with different behavioral repertoires. Queens and males are produced by colonies separately from the production of workers. The germ plasm analogy has been used persistently in building metaphorical arguments for the superorganism status of insect societies and continues today.

5.3 Termite Superorganisms

The idea of an insect colony as an organism found its way into the world of termite research with the work of Albert Emerson. Emerson was a professor of zoology at the University of Chicago and had a longtime affiliation with the American Museum of Natural History in New York. He was a close colleague of Wheeler, the two of them staking claims as the eminent ant and termite biologists of their time. Emerson embraced the superorganism metaphor but did little more than build analogies from the termites onto the concept.

The concept of insect societies as organisms reached a point of near absurdity with the writings of the South African lawyer, poet, and biologist Eugène Marais. Marais was a careful observer of animal behavior and wrote two books on animal behavior, one on termites (1937) and the other on baboons (published posthumously in 1969), exploring their psychology in search of evidence that they have souls. Even though he didn't use the term *superorganism* or directly recognize the work of Wheeler, his description of the mound-building termites of the South African veld was replete with analogies drawn from Wheeler's concept. Termite colonies are real organisms because, like us, they have a skin that marks the boundaries of their individuality, the mud walls of the nest, and, when damaged, is repaired from the inside, again like our skin is repaired. Repairs involve two kinds of workers, blood cells. One type of worker transports earth and water to the damaged site through passages within the mud structure, veins, while the other type, the soldiers, like our white blood cells, protect the wound against potential intruders.

Termite colonies, like us, have a "soul"; therefore, they're organisms. For Marais, a soul is a psyche, a goal-directed drive that comes from within that results in some kind of observable movement. A rock doesn't have a soul because only external forces work on it. A climbing plant does because it puts out tendrils and can grow toward a light source. Its movements come from within. We have a soul, but so do our individual organs; they function toward some goal of maintaining our bodies, but so do the individual cells of our body work for the organs. We're composed of a hierarchy of parts, each with a soul functioning for the human organism, all of which are controlled by the brain. A termite colony also has a group soul, beyond the individual souls of the workers, an organismic soul. Marais believed that individual workers have no instincts; all of their behavior is controlled by the colonial brain, the queen. In some way, she directs the activities of the colony toward a goal of nest building, defense, and colony nutrition—a superorganism.

5.4 The Spirit of the Hive

Marais struggled to explain how colony-level activities happen in a termite nest, how each individual, which he believed had no inherited instincts and no experience-based behavior, managed to engage in coordinated activities. He invoked a mysterious power of the queen to control the activities of the workers, just as our brains control the activities of each cell and organ of the body, which of course isn't the case. But this reveals the second dilemma of the superorganism, the first being how they evolve by Darwinian natural selection. Thirty-six years earlier (1901), Nobel laureate poet, playwright, and author Maurice Maeterlinck wrote a wonderfully romantic book on honey bees called *The Life of the Bee*. Maeterlinck brought to life the social activities of bees during the annual cycle of a colony. He recognized that their coordinated behavior seemed to be controlled by some force with an overall plan. But, unlike Marais, Maeterlinck didn't see the queen as the brain; instead, he resorted to mysticism to explain:

> She is not the queen in the sense in which men use the word. She issues no orders; she obeys, as meekly as the humblest of her subjects, the masked power, sovereignly wise, that for the present, and till we attempt to locate it, we will term the "spirit of the hive." (pp. 38–39; English translation 1913)

Neither the queen brain nor the spirit of the hive concept helps us understand complex social behavior. Darwin's explanation stands with only a need for more detail, detail that was provided 130 years later with the rise of complex systems analysis and studies of "emergent" social behavior (Chapter 7).

5.5 Decline of the Superorganism

After Wheeler, the superorganism did little to direct research and stimulated little in the way of new approaches to understanding social evolution. The superorganism remained little more than a metaphor, one that had limited use to biologists and fell out of fashion. Science had progressively become more reductionist in its approach. Holistic approaches, like those associated with natural history studies and superorganisms, had been replaced by more focused studies of mechanisms of behavior and physiology. The view from which science operated went from 10,000 meters to one centimeter. Instead of looking at the contour of the beach from above, scientists were studying the grains of sand, one by one. In 1971, Edward O. Wilson, another distinguished Harvard ant biologist, declared it dead in his book *The Insect Societies*. The metaphorical superorganism was built by naturalists with holistic views of the world, but science was occupied by reductionists looking for specific hypotheses to test with the latest research tools available. Those tools were used in studies of genetics, behavioral mechanisms, and physiology. Funding agencies, such as the National Science Foundation of the United States, wanted to fund hypothesis-based research; descriptive science was out. Universities wanted to hire new biologists who could get funding and would only tenure those that were successful, a consequence of their growing dependence on the overhead they received from the grants their faculty obtained. Overhead is a tax that the university takes from grants to build its research infrastructure, which enables it to get more grants.

The superorganism was spiraling down; it needed something more than the metaphor it had become. Attempts in the 1980s, and subsequently, to resurrect the superorganism and to extend the concept to all kinds of societies, including humans; ecosystems, like ponds; and even to the entire planet ended up confusing many practitioners in the field, like me. This occurred because of the repeated attempt to stray from Wheeler's essential criteria and view of superorganisms as true organisms.

5.6 The Superorganism Resurrected

The superorganism metaphor was reborn in the late 1980s with an article by David Sloan Wilson and Elliot Sober, a biologist and a philosopher. They proposed that superorganisms do exist but that they have very strict properties. First of all, they evolve by selection on groups, not individuals. This isn't fundamentally different from Darwin's theory of social insect evolution. However, selection on groups will normally fail to establish stable, enforceable social contracts without the additional help of inclusive fitness, which depends on groups composed of relatives. The problems were demonstrated in the cartoon in Figure 4.1 and the discussion of inclusive fitness in Chapter 4. But they also added the constraint that there can be no conflict or competition for reproduction among the parts, the workers in insect societies. In their view, group selection resulted in the total elimination of all reproductive competition below the level of the group. The view of no reproductive competition has been embraced by many social insect biologists, and the search for conflict and reproductive competition within societies has become a major research enterprise in insect sociobiology.

5.7 The Honey Bee as a Wheeler Superorganism

It's easy to paint a view of the honey bee as a Wheeler superorganism. The queen and the sperm cells she stores are the reproductive tissues, ovaries and testes. The workers are like soma, the body cells; they are unmated and normally functionally sterile. When a worker emerges from her cell, 21 days after the queen laid the egg, she enters a workforce of maybe 40,000. The workforce is differentiated by function and position in the nest, a division of labor analogous to cell types. They interact in integrated systems: larvae, nurse bees, food handlers and processors, nest builders, defenders, and foragers engaged in acquiring nutritional resources for colony reproduction. One can see the similarity of organ systems interacting to maintain fundamental body functions. The activities of bees within a cohort constitute a behavioral collective that's based on location and behavior, like differentiation of tissues in organ systems. The overall changes in tasks with age is called *age*, or *temporal, polyethism* and corresponds to a centripetal movement of bees from the center of the nest to the periphery, then outside. Tasks associated with reproduction, defense, and nutrition are compartmentalized to some extent into systems but interact with other such systems, as do organs. For example, nurse bees engaged in caring for and feeding larvae interact with the foragers indirectly by consuming stored pollen. A honey bee superorganism has internal communication similar to hormones and signaling peptides in us. The accumulation of stored pollen inhibits pollen foraging, similar to a state of satiation. As the nurse bees consume the pollen, stores are reduced, reducing the inhibition, hunger. The larvae are another part of the system by providing the stimulation for pollen foraging by producing chemicals called *brood pheromones*, analogous to major hormones that affect behavior. They are the final endpoint of the pollen collected. When there are many larvae, they consume the pollen faster and produce more "feed me" pheromone, resulting in an increase in pollen foraging. Older house bees, before they initiate for-aging, receive incoming nectar from foragers (sugar transport and storage). They take the nectar to storage areas of the nest, where they pass it off to bees that process the nectar. Nectar processors sit on the combs extruding the nectar from their mouthparts while it evaporates to the consistency of honey. They also add enzymes that break down the sucrose in the nectar to the component parts of fructose and glucose (sugar metabolism and fat and glycogen storage). The foragers adjust their foraging and recruiting based on the availability of receiver bees, who in turn are dependent on the rate at which they can pass their loads to nectar processors or place it into storage cells (feedback signals of satiation).

Similar systems exist for defensive behavior. Older house-aged bees guard the entrance of the nest, maintaining the social integrity of the colony. When an intruder attempts to enter, they alert the other bees in the nest to help in the colony defense by releasing a chemical called *alarm pheromone* (like adrenalin) that stimulates a defensive stinging response in other bees nearby. One group in particular, the soldier bees, respond to the stimulus. Soldier bees are a group that are older than the guards but forage less than others of their age. Defense against diseases and parasites also involve bees of different age-task groups (antibodies and white blood cells). Nurse bees have anatom-ical mechanisms to filter bacterial spores from food that's fed to larvae and produce antibacterial substances to protect them. Nurse-aged bees can detect diseased larvae and remove them from the nest. Older, middle-aged, bees remove dead bodies from the nest, thereby removing potential pathogens (phagocytosis), while bees engaged in nest construction and maintenance coat the inside of the nest with antibiotic materials collected from plants that seal the nest, like a wound-healing response.

Honey bees are organized into coordinated groups working together in sets of functions, like organ systems of higher individual animals. The nest itself and the colony that it contains is

a complete, definitely coordinated and therefore individualized system of activities, which are primarily directed to obtaining and assimilating substances from an environment, to producing other similar systems, known as offspring, and to protecting the system itself and usually also its offspring from disturbances emanating from the environment.

. . . it behaves as a unitary whole, maintaining its identity in space, resisting dissolution and, as a general rule, any fission with other colonies of the same or alien species. This resistance is manifested in the fierce defensive and offensive cöoperation of the colonials personnel. (Wheeler, 1911, pp. 308, 310)

5.8 Parting View

Insect societies exist in many forms. Reproductive conflict is rampant in many, cryptic in others, and perhaps missing in some, such as those species of ants where the workers no longer have ovaries and have lost all hope of personal reproduction. In others, like the honey bee, it's conditional on the life history of the colony. Can an insect society qualify as a superorganism during one life stage but not others? In Chapter 6, I present a kaleidoscopic view of honey bee societies and show how difficult it is to put a superorganism label on them or to reject it. I think the honey bee sits at the pinnacle of sociality. Does it matter what we call it? Probably not, but for the rest of this book, I'll adopt the Wheeler view of honey bees as superorganisms.

6

Reproductive Competition

There are more things in heaven and earth,
Horatio,
Than are dreamt of in your philosophy.

William Shakespeare, *Hamlet* (1603)

Sandy Mitchell, a professor of philosophy at the University of Pittsburgh, and I started a review of the superorganism with this quote from Shakespeare. We believed it captured what was missing in the attempts to place the vast array of colony organization into distinct categorical boxes. The resurrected view of the superorganism described at the end of the previous chapter is very narrow and idiosyncratic in that it defines something that's certainly rare in the world of insect societies and may fail to exist at all in the most strict interpretation. And yet, social insects are held up as the prime examples of what superorganisms are supposed to be.

I became acutely aware of the vast diversity of colony organization in bees many years ago when I was working in Uberlandia, Brazil, with Professor Warwick Kerr. I have known Kerr for many years and count him as one of the most caring and dedicated people I have known. One day he announced to me that we were going to travel to Viçosa, Brazil, to visit the bee research institute at the Federal University there. The driver called for us early in the morning; he had driven overnight from Viçosa, dressed in his uniform, California Highway Patrol–style sunglasses and driving gloves. We piled into his VW Golf, I folded myself into the back seat, and away we went, traversing the 500 kilometers (km) of winding roads, sometimes mountainous, at speeds at times of 160 km/hour (100 mph)! I was petrified the whole way.

When we arrived 5 or 6 hours later, we were met by Professor Lúcio Antônio de Oliveira Campos, the head of the bee institute at the university. He and Kerr have a long and intimate history. During the days of the military junta in Brazil (1964–1985), free speech was suspended. Students and university faculty who protested against the government were arrested and tortured, and many disappeared. Professor Kerr was jailed and "interrogated" twice! Think of the time in the United States, the late 1960s and early 1970s; students were actively engaged in political demonstrations, wanting their voices to be heard. The same was occurring in universities around the world, including Brazil. In 1970 or 1971 (exact date has faded), Professor Campos was arrested and taken to the police station for "questioning," at that time a prelude to disappearing, forever. Professor Kerr organized a vigil, a 24-hour watch over the police station, until Campos was released. This is a practice used by Amnesty International to make governments and police accountable for the political prisoners they take into custody. Professor Campos was tortured, then released. He credits Kerr with saving his life. He showed me the colonies in his amazing apiary of stingless bees, a cornucopia of social diversity.

The apiary had many different species of stingless bees, highly eusocial cousins of the honey bee that don't have stingers. Few of them build colonies containing nearly the numbers of workers as the honey bee, but some of them are equally defensive, using mass attacks with sharp, biting mandibles and sometimes caustic secretions that cause severe blistering of the skin. We went through boxes containing bees, some that were larger than our honey bee and some that resembled fruit flies with nests that could fit in a match box. Most typically have one queen; however, at least one species normally has two. In some species, laying workers normally lay the eggs that become males; in others, queens are normally the only egg layers. Competition and cannibalism among larvae determine who becomes a queen in some but not others. Each species' nest was different, with different ways of constructing comb, and many with no comb at all, just urn-like vessels for food storage and spherical orbs for nurseries connected by strings of wax suspending them in three dimensions, like a futuristic city from the early 1960s cartoon, *The Jetsons*.

In what follows, I present the diverse states through which honey bee colonies pass during their life history, the many kinds of reproductive states, and the competition that occurs, a kaleidoscopic view. This is a view that differs from the 10,000-meter (m) holistic view of the species-typical colony on which classifications are usually based and instead recognizes the incredible plasticity of individuals and the life histories of their colonies.

6.1   Colony Cycle of the Honey Bee

Social-organizational states of honey bee colonies change during different stages of their colony cycle. To examine the different social states of the honey bee, we need to define a honey bee colony life cycle. We need a starting point because it's seasonally circular and because they continually pass through different states. Let's start in the spring with a new virgin queen inheriting a nest. Her mother just left with a swarm and about half of her sisters, the continuation of her colony, leaving her daughter to form a new colony with the nest and remaining workers. At this point, the home nest may consist of 10,000 to 20,000 workers, her sisters, maybe a few hundred drones, brothers, and the new queen. At the time of honey bee colony division, there are usually several virgin queens remaining in the nest that will fight to the death for reproductive dominance, reproductive competition. One queen emerges the victor and heads a nest full of sisters sharing in cooperative brood rearing but no overlapping generations—so, by definition, not at the pinnacle of sociality. The pinnacle is occupied by colonies that are called *eusocial*, defined as sharing a nest, having a reproductive division of labor (queen and workers), having overlapping generations (mother and her daughters), and engaging in cooperative brood care. At this point in the colony cycle all individuals constitute a single generation, no overlap. This state has been called *semisocial* because all colony members are of the same generation. The colony will remain semisocial until the new queen has mated and laid eggs and a new generation develops. Then, the colony is a mixture of her sisters and daughters. It will require 4 to 6 weeks for her sister workers to work themselves to death, leaving a colony composed of the new queen and her daughters, a eusocial colony.

The other half of the original colony contains the old queen and her colony. It assumed the risk; it must find a new place to live, clean it, build combs, produce more workers to expand the population, and provision it for winter. It remains eusocial. However, many colonies supersede the old queen after swarming and the new location is established. It's the optimal time to do it. It's late spring or early summer, drones are still flying who can mate with the replacement queen, and enough time is left in the season to build a new colony. All queens must die at some time; it's better to replace them when time is optimal rather than have them die at a time when they can't be successfully replaced. When a colony fails to replace its queen, it becomes queenless. There are different potential fates for a queenless colony that depend on the subspecies of the bee and its success in replacing the lost queen.

6.2 Reproductive Competition in the Superorganism

Honey bee societies are rich in reproductive competition and conflict, depending on their current social states. In the following sections, I present the amazingly complex web of competitive interactions.

Competition among Males

In temperate climate–adapted honey bees, such as are found in Europe and those introduced and distributed throughout North America for commercial beekeeping, the colony re-achieves a highly eusocial status by late spring or early summer. At this time, colonies have many drones, sons of the departed mother queen, produced to coincide with the need for mates of new queens produced during the spring swarm season, each queen using up and discarding 20 or so males. The poor drones just mate once and die in the act. In central California, the peak time for raising drones is mid-April to mid-May, corresponding to the annual spring swarm season that peaks in May. The peak of drone brood production occurs about 30 days before the peak of swarm season, the correct amount of time for the drones to emerge, mature, and be available to mate with the new queens produced. Drones aren't part of the cooperative division of labor that defines the superorganism. They don't care for brood or construct, defend, or maintain the nest. And they don't forage. They consume resources, begging them from their sisters who feed them mouth to mouth, a behavior called *trophallaxis*, and fly out from the nest every afternoon seeking flying virgin queens to mate with. If successful, they die in the act and fall to the ground, food for ant superorganisms. If unsuccessful, they return to their home nest and beg food again until the next day. Males within the nest compete for resources begged from their sisters, then compete among themselves and thousands of other drones attempting to mate with virgin queens within a 5-km radius, 78 km^2—reproductive competition.

Competition Between Workers and the Queen

Worker honey bees aren't anatomically sterile as are some ants. They each have two ovaries that are capable of producing eggs. The honey bee ovary consists of filamentous tubules called *ovarioles* that are elongated and serve as individual factories for egg production. Queens have usually between 280 and 320 ovarioles, while workers have four to 12. Based on an average of 1,500 eggs laid per day for a queen during the spring and summer and an average of 300 ovarioles, we can estimate that each ovariole is capable of producing around five eggs per day. Therefore, assuming a worker ovariole is capable of the same, probably not a valid assumption, a typical worker with about eight ovarioles is capable of laying a maximum of about 40 eggs per day. However, workers are chemically sterilized. Production of eggs by worker ovaries is inhibited by chemicals, pheromones, produced by larvae and the queen. I find it interesting that the queen doesn't inhibit herself, nor is she inhibited by the brood, so there's an inherent physiological mechanism to avoid it. Workers must in some way cooperate in their own sterilization by responding to the chemical signals from the queen and brood. But not all workers are inhibited to the same degree. As the colony grows during the spring and summer, the chemicals produced by the queen and larvae must be distributed to more and more workers. For some, the chemical titer isn't high enough to completely inhibit the development of their ovaries and egg production. Workers aren't mated, so the eggs they lay turn into drones.

When I was a postdoctoral researcher at the US Department of Agriculture's honey bee research laboratory in Madison, Wisconsin, I decided to try to get an idea of how many males in colonies are produced by egg-laying workers, rather than queens. Early honey bee geneticists warned that some of the drones that might be collected from colonies for instrumental insemination of queens used in breeding programs might actually be offspring of workers and not from the queens selected for breeding. In the apiary, we had colonies that had queens mated to drones that carried recessive genes for different eye color mutations. The queen did not carry the eye mutation genes; therefore, her worker offspring had normal eye color, black, but were heterozygous for the mutation. Half of the males derived from the worker-laid eggs should express the eye mutation; the other half should be normal and indistinguishable from those of the queen.

Out of 13 colonies examined, five (38%) produced at least some drones with colored eyes, sons of workers. Altogether, drones from workers were rare. In queenright colonies 7% to 45% of workers may have ovaries developed sufficiently to lay eggs, depending on the race of bee and the season. Ben Oldroyd in Australia, however, identified colonies that produced significant numbers of worker-laid eggs and bred them, increasing the frequency of worker-laid eggs. Normally, worker-laid eggs are detected by other workers and eaten. The result is that adult drones derived from laying workers are normally very rare in colonies that have healthy, functional queens, about 0.1%. Nonetheless, the reproductive conflict between queens, egg-laying workers, and workers that consume the worker-laid eggs disqualifies the colony from superorganism status under the strict resurrected view.

Competition Following the Loss of the Queen

The ultimate fate of all colonies is the loss of the queen. Queen loss is a branch point in the development of a colony, leading to five different possible colony states.

Queen Is Replaced by Supersedure

Queens are replaced, superseded, through a coordinated process when they age. This isn't the same for all queens. Some are superseded within the same season they are produced; others may live for 2 or more years. The average is probably less than 1 year. In commercial colonies in California and Mexico, queens on average survive about 9 months. That means that some are superseded soon after they initiate a colony; others survive 1 year or more. As a consequence, at any given time there are colonies existing in various states of social development within a population.

The exact cues and signals used by bees to coordinate a process of replacing an aging or declining queen aren't fully known or understood. But it begins with workers preparing the wax bases of a few queen cells. The queen then lays an egg in each prepared cell, usually just a few, making her a co-conspirator of her own demise. The workers raise the larvae, and a few queens mature. The virgins are culled to just one remaining through a process that includes one-on-one combat and unbiased assassination by the workers. The workers take no sides with respect to whether the queen is a sister from the same drone father as they or a half-sister: no nepotism. The mother queen remains in the nest until the new victorious virgin has mated and matured to laying eggs. Then, they may both remain actively laying eggs for a period of days or longer. Eventually, the old queen is killed and cast out, a plot befitting a television drama.

When I was a graduate student, I was skeptical of everything. I was taught the dogma that honey bee colonies only have one queen. But I noticed that when we look for a queen in the colony, we stop when we find one, like the question we all ask ourselves "Why do I always find my missing sock in the last place I look?" We don't keep looking to see if there might be two. So one day I went through all of the colonies in my apiary, about 25. I searched them thoroughly, looking for additional queens. Three had two queens! I estimated 10% to 15%. These queens appeared to be active egg layers, fat, walking on the comb, inspecting cells. Perhaps these were colonies undergoing supersedure? I couldn't determine that. But, at least at this point in time, the colony had its reproductive interests going through more than one individual but was still highly eusocial and would still qualify as a superorganism, by Wheeler, but not as a superorganism by the tougher criterion of "no reproductive competition."

Queen Dies and Colony Replaces Her

Queens die. Like us, she gets old, gets diseases, or, in an apiary, falls off the comb when a careless beekeeper is looking for her and she gets stepped on. When I was a postdoctoral researcher, I did a project where I was studying how queen bees use the sperm of the many drones they store in their spermatheca. Queens were inseminated with the sperm from each of six different drones that carried genetic mutations that we could use to distinguish the father of each worker produced in the colony. The queens were put in colonies from which I sampled newly emerging workers, determined which mutations they carried, and assigned them to a drone father. I was doing my normal colony inspections one day, looking for the queens to make sure they were there (they were individually marked with colored plastic disks with unique numbers), and couldn't find one of the queens. I looked and looked, going through the hive frame by frame. At that time, I was an avid runner and always wore waffle-soled running shoes. I looked down and found the queen stuck to the waffles of my shoe. I was sick. But worker bees have no emotions, as far as we know; and instead, if there are young larvae in the nest, the workers select a few of them within hours and begin a process of raising new queens.

Why some larvae are raised to be queens and others rejected isn't well understood. But through some process that involves larvae signaling nurse bees, some are selected. There's probably some kind of competition, like larvae shouting "choose me" with chemicals, because studies have shown that larvae from some genetic sources are more likely to be selected than those from others. The payoff for being chosen is that you inherit the nest and become the mother of all reproductive males and queens produced by the colony.

If the colony is successful at producing a replacement queen that mates and begins egg laying, it returns to the highest, "typical," social state, eusocial. If not, the colony becomes "hopelessly queenless," hopeless because by the time they attempt and fail to produce a queen, all of the remaining larvae are too old to become queens. Larvae in worker cells, destined to become workers, can be developmentally redirected into queens up until about 2.5 to 3 days old. It takes 3 days for an egg to hatch; therefore, the window of time for a colony to respond to being without a queen is only about 5.5 to 6.5 days. They get only one chance.

Terminal Laying Worker State
In the absence of larvae and the queen, workers undergo reproductive development. Their ovaries swell and become activated to receive proteins from the blood that are converted into egg yolk. Eggs then begin to develop in their ovaries, and they initiate egg-laying behavior. This state is often considered to be nothing more than the collapse of the colony, a condition of total chaos, a consequence of the loss of control systems responsible for social order, a corruption. This attitude of laying worker colonies not being "real" honey bee colonies has led to relatively little research on them. But it's much more than that. It isn't rare. It's the fate of nearly all colonies that lose their queens during times when there are no eggs and larvae available, which can

be several months of broodlessness experienced by colonies in the northern temperate regions. In California and Wisconsin, I estimated that we lost about 10% of the queens during the winter, even with careful management. When the colonies were examined in the late winter or early spring, some were full of egg-laying workers.

The time it takes to have egg-laying workers is different among the different subspecies of honey bees. Subspecies are also referred to as "races" in the beekeeping world. For example, the main bee used for hobby and commercial beekeeping in the United States is known as the Italian bee. It's of the Italian race. But its subspecies name is *Apis mellifera ligustica*. Another popular bee is the carniolan race. It's *Apis mellifera carnica*. The time it takes different subspecies to develop into laying worker colonies varies from 3 to more than 30 days. At first just a few eggs are laid by workers, then over a period of days the number of eggs observed within cells dramatically increases. The number of workers with developed ovaries increases until there may be hundreds. Workers can be observed backed into cells depositing eggs. Queens are longer than workers, and when laying an egg, their head and much of the thorax is out of the cell. The abdomen is agile and locates and places the egg centrally at the bottom of the cell before the queen withdraws. Workers have a shorter abdomen and body and back into the cells up to their necks, only the head protruding. Their wings are poorly placed for egg laying and often get caught on the edge of the cell and fold back over their head. Workers' abdomens are less agile and place the egg with less precision at the bottom. When given a choice of depositing an egg in a drone cell or a worker cell, laying workers show a bias for drone cells. This is different from mated queens, even virgin queens. Queens normally show a bias for worker cells.

Beekeepers identify laying worker colonies by the presence of the typical torpedo-shaped cappings on the worker-sized brood cells and the egg-laying pattern they observe. Laying worker colonies in typical human-made hives have multiple eggs deposited in the same worker-sized cells. The eggs appear poorly placed in the bottom of the cell, often on their sides rather than upright and pert, projecting from the bottom of the cell. But this is an artifact of our beekeeping. We restrict the amount of drone comb in colonies; we want our bees to raise more hard-working workers and fewer lazy, food-consuming drones. Thus, when a colony is queenless and laying workers begin their activities, there aren't enough drone cells available for their eggs, so they lay in worker cells that are centrally located in the brood nest. At first, you observe just one egg per cell, each placed at the bottom. Over time, the workers lay additional eggs in the same cells due to an insufficient number of central cells. When a worker backs into a cell that already has an egg, she often bumps it with her abdomen, knocking it over so that it looks poorly placed.

When unfertilized eggs are deposited in worker cells, worker-sized drones are produced, about half the size of a normal drone raised in a drone cell, with fewer spermatozoa. Competition among drones is intense at congregation areas where thousands may assemble and chase virgin queens on mating flights. The numerical sex ratio of a honey bee population may be as high as 10,000 drones for every queen taking a mating flight over the season. This ratio assumes that colonies with queens produce 20,000 drones over a season, which they can do, and that each colony produces two successful queens, one that remains when the mother swarms and the other when the mother is superseded. The sperm from each of the 20 lucky enough to mate compete for getting to the sperm-storage organ of the queen and finding a place inside. Each male can deposit 6 to 8 million sperm in the queen, up to 160 million combined. But the queen stores only about the amount one male deposits, about 5.5 million. The more sperm a male deposits, the better his chance that one of his sperm eventually gets to fertilize an egg that becomes a reproductive queen, the greater the chance that his own mother, queen or laying worker, will be reproductively successful through him.

Assuming there are two queens produced per colony, only about 40 of 20,000 drones on average will actually mate with a queen, one in 500, and of those, four (two per queen) will fertilize an egg that becomes a queen. Overall, one in 5,000 drones successfully reproduces. Only the biggest, fastest, and most keenly aware drones can hope to be successful. (Fertilizing eggs that develop into workers isn't reproduction, except in the extremely rare case where a worker becomes a successful laying worker.)

Natural selection has resulted in egg-laying behavior that increases the successful reproduction of workers: lay eggs in drone cells and get large, competitive drones. The total number of full-sized drones produced by the collective egg laying of the colony can be more than 6,000 before the social system of the colony declines and is no longer able to raise brood, a potential of eight males that mate, a probability that 0.8 will fertilize a queen-destined egg—the colony's successful reproductive output. Some workers have a chance to have grandchildren.

Laying worker colonies have a division of labor within the nest that's different from queenright colonies. The division of labor is based on competition and cooperation. As in the paper wasps, workers are very antagonistic toward other workers that lay eggs, sometimes resulting in death. As the ovaries develop, the adult worker passes into different roles, more competition and less cooperation. Bees with the most developed ovaries engage in egg-laying behavior, though they may also engage in other activities, including pollen foraging. There are two stages of this behavior. The first is when workers initiate egg-laying behavior without actually depositing an egg. They back into a cell, as if they are depositing an egg, and pull out leaving nothing behind to show for their effort. Others actually lay eggs. Ovary development doesn't seem to differ between these two groups. Some bees continue to provide care for the larvae, and others continue to forage, for a while. They have ovaries in less developed states; they aren't making eggs, yet. The workers with the least developed ovaries engage in egg eating, called *oophagy* in entomology-speak. In some cases, we have observed many eggs being laid followed by many egg eaters, resulting in no hatched eggs for several days; and, in extreme cases, the competition continues until the colony never produces any adult drones. Some workers will continue to construct comb for a while, but even comb construction is in their own selfish interests. They construct mostly drone comb, future repositories of their own eggs.

Egg layers are competing for a resource, the care of the nurse bees and the foraging efforts of the foragers, before the social system fails and collapses into anarchy and chaos. They depend on them to get their offspring reared. But, at the same time, the nurses and foragers are undergoing changes of reproductive status and moving toward reproduction, leaving many things within the nest undone. As a consequence, the window of time for laying eggs that will become adult drones is small. It's easy to see that rapid ovary development in an individual worker can be rewarded by producing more drone offspring. However, early egg layers are treated poorly; other bees are very aggressive toward them, and their eggs are eaten. But what do egg cannibals gain? Egg cannibals don't yet have ovaries capable of producing eggs. They develop more slowly. However, being hopeful reproductives, they reduce the future competition of their own eggs and larvae. In the process, they also consume eggs full of protein that they can recycle into eggs of their own when they begin. There's a limit to the amount of oophagy a colony can tolerate without a catastrophic reproductive failure.

In a sense, ovaries are the competitors in laying worker colonies; the workers just carry them around. The division of labor in laying worker colonies is linked directly to ovary development. Workers with more ovarioles develop their ovaries faster and assume egg-laying roles sooner. Behavioral repertoires of workers (and queens) are linked to the reproductive signaling networks of the ovaries, changing with ovary state. This is true in most animals, including us. Behavior changes with reproductive status. We observe many changes in human behavior throughout puberty and even with changes of ovary status with the menstrual cycles of women.

The normal endpoint for a laying worker colony is the loss of social organization and slow death of all of the workers within it. After the initially synchronized burst of successful drone rearing, regulatory systems give way, and the colony collapses into competitive chaos. No longer a functional superorganism, eggs are laid that only get eaten. Larvae from those that may hatch aren't cared for. Some bees continue to forage, disconnected from the asocial network; but most just consume the stores and engage in hapless efforts to reproduce. The colony dwindles for months until the last solitary bee finally dies.

False Queens

Gene Robinson and I were studying queenless colonies when we were at Ohio State University. One morning, Gene, a keen observer, had been out inspecting the colonies and came into the lab excited and proclaimed to me that one of the colonies we had set up in our study had a false queen. Neither of us had ever seen one before; they aren't common in laying worker colonies. On further inspection, he found four false queens in three of the 15 queenless colonies from our study. Two colonies had one; the other had two. False queens stood out from the other several thousand bees by having a distinct retinue of workers surrounding them, caring for them like queens of queenright colonies. The false queens produced pheromones like those of queens but in smaller amounts. When we dissected them and compared them to the other workers in the colony, we found that their ovaries were no more developed than the others. The other workers in the colony had developed or developing ovaries; the false queens were not suppressing ovary development, only egg laying. They suppressed the egg-laying behavior of the other workers even though they themselves laid few or no eggs, an interesting case of spite, perhaps, where competition for egg laying resulted in the failure of the colony to produce any successful drones—failed reproduction.

False queens are rare in queenless colonies from European races of bees, but the trait is heritable; it runs in family lineages, as was the case for our four queens. They had the same mother and father. The trait appears to be maladaptive in the context we were testing, leading to reproductive failure; however, if coupled with other traits, such as enhanced ovary development and egg production, and the ability to produce female offspring from unfertilized eggs, it could be a trait that could restore a hopelessly queenless colony to a queenright, highly eusocial society, as often happens in the Cape honey bee.

Queenright Again

In very rare circumstances, a laying worker colony might requeen itself. Worker honey bees don't mate; if they tried, it would be a bizarre experience considering that the genitalia of the drones are probably the size of a worker. The spermatheca of a worker is usually very small compared to a queen and doesn't have the accessory glands necessary to keep the sperm viable inside. Their reproductive plumbing isn't engineered for the task of fertilizing eggs. But they can produce diploid female offspring through a special kind of parthenogenesis called *thelytoky*. Normally, unfertilized eggs of the queen and egg-laying workers develop into males, called *arrhenotoky*. Arrhenotokous or haploid males have just one set of chromosomes that they inherit from their queen or worker mother. Eggs and sperm are derived from special germ tissues, the ovaries and testes. Primary sex cells that were sequestered and protected early in embryo development undergo division to produce the gametes, the egg and sperm cells, through a process of meiosis that results in a reduction of the number of chromosomes contained in the cells to a single set that consists of genes inherited from both parents. Honey bees have a base number, N, of 16 chromosomes. Nuclei of gametes, eggs and sperm, contain the base number of 16. Males have no contributions of a father's sperm, so they begin development only with the chromosomes they inherited from their mother, a total of 16. They are haploid, N. Females are diploid, 2N, derived from the fusion of an egg and a sperm. They begin development with a total of 32 chromosomes, though during development individual cells may replicate chromosomes to higher values of N to perform specific functions within some tissues. This is the basic haplodiploid sex determination found throughout the order Hymenoptera that includes the bees, wasps, and ants.

During the process of oogenesis, egg making, the parent cell giving rise to the egg undergoes a process of meiosis. The chromosomes from each parent find their counterparts and pair. Then they duplicate, making a total of 4N, or 64 chromosomes. Next, they exchange DNA between chromosomes from different parents, then separate into four sets of N chromosomes. Each set of chromosomes becomes a kind of nucleus with the chromosomes contained within a membrane, one of which becomes the pronucleus and the others, polar bodies. If the queen releases sperm from her spermatheca to fertilize the egg, a few sperm enter the egg through the open matrix of the micropyle at the end of the egg and race to be first to find the pronucleus. The nuclear contents of the first one there, the 16 chromosomes contained in the sperm, fuse with the 16 chromosomes of one of the nuclei, the pronucleus, and form a zygote, 2N, 32 chromosomes. The other three nuclei, the polar bodies, disintegrate; they contribute nothing to the new life now contained in the egg. If no sperm cell fuses with the pronucleus within a short time, it will undergo development on its own, resulting in cell division and an individual with just one set N, of chromosomes, 16. This will be a male.

However, in rare cases, less than 1%, an unfertilized egg laid by a worker or a queen will undergo a different process. Two of the nuclei contained in the unfertilized egg fuse with each other, instead of having a sperm to fuse with, and form a zygote. The zygote is 2N; contains two sets of chromosomes, both from the same worker or queen; and develops as a female. This is called *automictic thelytoky*, a kind of parthenogenesis. Even more rarely, it's possible that the rare laying worker that produces a thelytokous egg deposits it in a queen cell, and it's raised to be a queen, thereby requeening the "hopelessly queenless" colony. It can happen. There are reports of it, though it's very unlikely.

When I was a young researcher, I was fascinated by laying workers, I guess because it was a countercultural area of research. Few cared about them, and most treated them as anomalies. But I always had a few colonies around that I had dequeened and had under study. I had one colony with a wild-type queen mated (by instrument) to a cordovan drone. Cordovan is a recessive mutant gene that turns the black parts of the body, like the head, thorax, legs, antennae, and black stripes on the abdomen, to a brown color that sometimes is similar to cordovan shoes. It only changes the color when homozygous, meaning the individual inherited two copies of the mutant gene, one from each parent. The workers in this colony were heterozygous, meaning that the set of chromosomes they got from their mother had the wild-type gene for integument color but the set they inherited from their father had the cordovan mutant gene. They did not have the cordovan trait; all looked wild-type. The same would be true for any queens produced. During an examination of the hive, I discovered the colony had a cordovan queen and a queen cell that looked like a queen had emerged from it, a smoking gun. It couldn't have been produced by the previous queen because it was several weeks after she was removed, and she would have produced a wild-type daughter queen. There were no other colonies around that carried the cordovan gene, so it was unlikely that a virgin queen produced in one of them drifted accidentally into this hive. I deduced it could be a queen produced by thelytoky. Thelytokous workers could produce cordovan workers and queens.

6.3 Social Parasitism in the Cape Honey Bee

In the early 20th century, reports were coming out of South Africa about a strange honey bee where the workers laid eggs that became females, rather than males. This was met with disbelief in the bee journals until a series of papers were published supporting the findings of the beekeeper who initially reported it. In 1845, Johann Dzierzon, a priest from Silesia, a region of what's now part of Poland, published one of the most important discoveries in honey bee genetics, the fatherless origin of male honey bees. By the close of the 19th century, chromosomes had been discovered, followed early in the 20th century by a chromosome theory of inheritance. So by the second decade of the 20th century it was well known that workers are unable to lay fertilized eggs and that unfertilized eggs give rise to haploid male offspring. The reports from South Africa shook the very foundations of the nascent field of honey bee genetics.

(A not-so-short segue back to Dzierzon: When I was a graduate student at the University of California at Davis, I spent much of my time at the Bee Biology Laboratory, now the Harry H. Laidlaw Honey Bee Biology Facility. In the hallway was a glass-fronted cabinet that contained some history of apicultural science including a picture of Dzierzon wearing medals he had received from beekeeping organizations. One large one in particular caught my notice, pinned to the breast of his jacket. They looked like military medals. Dzierzon was revered as the father of honey bee genetics, but Professor Harry H. Laidlaw was the father of modern honey bee breeding and genetics, having developed the methods for instrumentally inseminating honey bee queens, thereby controlling matings and opening the door for real genetic studies. Harry had his academic career at Davis and became my best friend and mentor for 25 years. The picture of Dzierzon occupied the same shelf as the collection of early instrumental insemination instruments that Harry invented.

In 2001, I was visiting the University of Würzburg for a month, working on a review article for the journal *Genetics*. We were living in a small 16th-century half-timbered house in a village on the Main River. One beautiful Sunday morning the owner of the house, Michael Zimmermann, knocked on the door and invited me outside into the sunlit courtyard that was surrounded by my house, his house, and the barn, for a little *frühschoppen*, literally "early drinking," a tradition in Germany. He brought out a bottle of Franconian wine and started telling me about his beekeeper father-in-law. His wife had a great, great, great . . . uncle who was a famous beekeeper. His name was Dzierzon. I was excited. The reason I was in Würzburg was to write a short history of the genetics of the Hymenoptera, ants, bees, and wasps. And at that very moment I was writing about Dzierzon. I ran inside and brought out my manuscript and the articles I had about him and showed them to Zimmermann. He arranged for us to visit the apiary of his father-in-law and look at his bees.

Hobby beekeeping in Germany is a lifestyle, a central part of who you are and what you do. Often, the apiary is contained adjacent to a small vacation home, a shed-like structure sitting on a small piece of land, usually less than an eighth of a hectare with a garden. Inside the structure may be a daybed, places to sit, and a table for lunch. We were invited for afternoon cake and coffee, another German tradition. After examining bees with his 80-year-old father-in-law, we sat down for cake and his mother-in-law brought out a small box that contained the medals of her great uncle, the very ones he wears in the picture in hallway of the bee lab at Davis! I was overwhelmed. His father-in-law gave me a gift of the very last bottle of his homemade honey wine, mead. I took it back to Davis and shared it and my extraordinary story with my lab. Zimmermann's father-in-law died shortly after, truly his last bottle.)

Little was done to further characterize and understand the Cape bee until the early 1960s when a South African apiculture doctoral student from Cornell University, R. H. Anderson, conducted experiments that greatly added to our understanding of their unusual biology. Research to date has given us a picture of a strange bee, one where the ideas of sociality and super-organism are stretched and broken. The Cape bee, *Apis mellifera capensis*, is the same species as the bees from Europe that are used throughout the United States for commercial and hobby beekeeping and the other honey bees that are native to Africa. It's found in a small part of the southern tip of South Africa, the fynbos, an area that's known for its unusual ecology and plant types. Stretching in a band 100 to 200 km wide across the southern tip, it's considered one of the six world floral kingdoms and contains more plant diversity than any other place on earth, including the Amazon rainforest. Of its more than 9,000 plant species, 6,200 are endemic, found nowhere else on earth. The Cape honey bee is restricted in its range to this small area but interbreeds freely on its edges with *Apis mellifera scutellata*, the African highland bee that was introduced into South America and gave rise to the Africanized, or killer, bees. But unlike the fiercely defensive bees that occupy most of southern Africa, the Cape bee is very docile.

The best-known characteristic of the Cape bee is its production of false queens in the absence of a queen and its ability to produce female offspring through thelytoky. As explained earlier in the sections "False Queens" and "Queenright Again," both of these traits are found in other races of bees, though rare. In the Cape bee, they are the norm, not the exception, with more than 99% of eggs laid by queenless *capensis* workers normally becoming females, workers and queens. Queens also have this ability and apparently can control whether the unfertilized egg develops into a male or a female. *Capensis* workers normally don't lay eggs in the presence of a *capensis* queen, except when colonies are preparing to swarm. Then they may lay eggs that become queens or workers.

Cape worker honey bees have adapted a parasitic lifestyle in parallel with a normal honey bee superorganism phenotype. Within their home range Cape workers normally don't reproduce in queenright colonies, though during preparations for swarming as many as 10% of the workers produced come from worker-laid eggs as do 40% to 60% of the queens produced for swarming. This is clearly very significant reproductive competition. When they lose their queen, laying workers develop quickly and begin producing clones of themselves. Individual clones, the descendants of single workers, may engage in clonal wars within the nest until a new queen or false queen(s) emerges and restores order.

On the edges of their distribution where they meet *A. m. scutellata*, things are different. Workers enter foreign colonies and begin reproducing. Each egg gives rise to a clone of the worker that laid it. If more than one worker enters a foreign colony, then multiple clones can exist, each one competing with the other for resources and with the queen and workers of the host colony. The clonal workers don't work. They simply lay eggs and consume resources. In time, they outnumber the original residents and consume all of their food, the resident queen dies (they kill her about 5 to 6 weeks after they invade), and the colony dies. But, before they all die, the *capensis* workers disperse, finding new colonies to invade.

Cape bees episodically wreak havoc on the beekeeping industry of South Africa. Catastrophes occur when Cape bees are accidentally or deliberately brought out of the fynbos into other beekeeping areas. Their clones quickly spread, killing off large numbers of *scutellata* colonies. *Scutellata*, though the source of Africanized bees, vilified throughout the Americas, is a very important and productive bee in Africa. The bee industry depends on it. One catastrophic invasion of commercial bees early in this century was found to be derived from a single worker estimated to have lived in the 1980s. This individual gave rise to a clonal population of billions of workers spread throughout countless colonies of *scutellata*.

Cape workers have other traits that assist them in their parasitic lifestyle. As larvae in *scutellata* colonies, they are able to beg more food resources from the *scutellata* nurses and grow faster, are larger, and have larger ovaries that contain more ovarioles (about 20 in some populations) than workers of other races (usually fewer than 12). This allows them to lay more eggs than other races of bees. Their eggs are accepted as if they were laid by a queen. Normally worker-laid eggs are detected and are removed and eaten by nurse bees, but *capensis* workers can mask their origin, having eggs indistinguishable from those of queens. They also activate their ovaries much faster in the absence of a queen, about 3 days compared to about 10 days for *scutellata*; and they live longer, 3 to 5 months compared to about 6 weeks during the active foraging season. They regularly develop false queens that produce queen pheromones from mandibular glands, as do rare individuals of other races. However, they also have tergal glands on the abdomen, like queens, another source of queen pheromone. Workers of other races don't. As a result, false *capensis* queens suppress egg laying and queen rearing in other workers, except not Cape workers. *Capensis* workers are immune to the suppressing pheromones of other workers, as well as queens and larvae. This gives them a significant advantage in competition with queens and workers in *scutellata* colonies, a kind of reproductive dominance. And queenless *capensis* workers, unlike their sister races of bees, construct worker comb in the nest. Queenless workers of other races of bees construct only, or primarily, drone comb.

The collection of traits was assembled over many generations by natural selection. Each of the traits exists independently in other races of bees, at low levels. The traits aren't controlled by a master gene or genes that turn on the whole collection, but each trait is controlled independently and varies genetically among workers. Selection on the Cape bees in the fynbos assembled the traits into a Cape honey bee genotype that produces a phenotype fit for the ecology of their habitat. The climate is Mediterranean, similar to that experienced by Italian bees, so that doesn't explain the shaping of the unique *capensis* laying worker. However, it's very windy, and queens must take mating flights in relatively high winds. *Capensis* queens have evolved to take mating flights in winds as strong as 14 m/second (50 km/hour) compared to 5 m/second (18 km/hour) for *scutellata* and other races of bees. And queen loss on mating flights is high, leaving many colonies hopelessly queenless. The suite of traits favoring worker reproduction may have evolved in response to queenlessness and offered a chance for colony survival and reproductive success. The suite of traits preadapted the bees for their parasitic lifestyle when in competition with other populations; nothing new was needed, just the opportunity (see Table 6.1 for a summary of possible states of honey bee colonies).

Table 6.1 Different Possible Social States of Honey Bee Colonies Based on the Reproductive Status of the Queen and Workers

Colony state	Reproductive status	Social conditions	Reproductive competition	Fate
Queenright	One laying queen	Brood present	No (assuming no competition among drones)	Perennial
	One laying queen	Laying workers present	Yes	Perennial
	One laying queen	No brood present post-swarming	None or very low	Perennial
	Parasitized by capensis	Female producing laying workers	Yes, intense	Death
	Two laying queens	Post-supersedure	Yes, between queens	Perennial
Queenless	Laying workers	Laying workers produce males	Yes, intense between workers based on physiology	Usually death but may rarely requeen
	Laying workers	Laying workers produce females	Yes, between clones	Death or requeen
	False queens	Male producing	Yes	Death
		Female producing	Yes	Death

6.4 Disease in the Superorganism

I have multiple myeloma cancer. Some of my bone marrow cells stopped listening to their neighbors and began dividing uncontrolled. Some years ago, hard to tell how many, I had something happen in the chromosomes of one single plasma cell (white blood cell), somewhere in my bone marrow. There's no single gene that causes this disease; there are many different kinds of changes, mutations, translocations of pieces of chromosomes, etc. that all result in the same disease phenotype in the end. Some they know and can detect, others not. Mine not. This plasma cell, because of this event, or more likely a series of cumulative changes, stopped listening to signals from its neighbors. It made new cells just like itself, clones that then made cells like themselves. First there was one, then two, four, eight, 16, 32, 64, 128, 256, until there were millions or billions, so many that they were crowding out the other healthy ones. The mutant clone cells don't contribute to the division of labor of bone marrow cells and, as a result, don't contribute positively to my overall health and well-being. Healthy plasma cells are white blood cells that make antibodies, each lineage of cell making one kind of antibody; perhaps mine—when it was healthy—made an antibody against a protein found on the surface of a bacterium that caused an ear infection when I was a child. Our body makes billions of them; they give us immunity. My cancerous plasma cell clone used to make an immunoglobulin G (IgG) antibody that was constructed from four protein chains produced by the cell, two pairs. One pair is twice the size of the other and is called the *heavy chain*, the other the *light chain*. Light chains come in two varieties, kappa and lambda. The four strands are assembled into one larger protein, the IgG. My cancer clone cells proliferate uncontrolled and produce an excess of the kappa light chains that get released into the blood, causing many problems. The clone also consumes energy resources and takes up space in the bone marrow, reducing the other antibodies I need to protect myself.

The cells in our body respond to control signals that are both global and local in scope. Hormones are secreted into the body and act everywhere with a wide variety of effects. For instance, thyroid hormones are produced in tiny amounts by the thyroid gland with broad effects on protein, carbohydrate, and fat metabolism as well as cell differentiation and gene expression, affecting every cell of the body. But cells also give and receive signals from other cells around them. Of the 37 trillion cells in the human body there are only about 200 different kinds of cell types, for example, erythrocyte (blood), osteoclast (bone), dendritic (lymphoid), or microglial (central nervous system). This is amazing considering we have around 20,000 genes, the activity states of which determine the type of cell. The number of possible states is staggering. No central control system could possibly be complex enough to orchestrate the development and physiology of our bodies. Cells divide, move, are induced by their environment (global and local signals), differentiate into specific cell types, and die. So, cells that occur in specific tissues, like the liver, get signals from others around them inducing them to differentiate into liver cells (hepatocytes) rather than kidney cells (glomerula). Cells also regulate each other, inhibiting division of cells (reproduction) when more of them aren't needed and stimulating division when needed. Cell-to-cell signaling over short or intermediate distances is through chemicals and physical contact.

The parallels with honey bee societies are obvious. Bees are born in the center of the nest, undifferentiated, and able to develop into different worker types with specialized physiologies and behavior, like stem cells. As they age, they move and are induced by their surroundings into different roles such as nurses, food processors, entrance guards, and foragers. The queen produces chemical pheromones, perhaps analogues of hormones, with broad effects that regulate the social tempo of colonies, stimulate foraging behavior, and inhibit the construction of queen cells and ovary development in workers. Young larvae also produce chemicals with similar effects as the queen and regulate the collection and storage of protein and carbohydrate in the superorganism "body." Adult workers produce 50 different known chemicals that affect the physiology and behavior of those around them. The physical interactions of nurse bees and foragers also affect the physiology of individuals, resulting in changes in life states of being a nurse or forager. Orientation pheromones result in the recruitment and movement of workers to specific locations, and alarm pheromones change the state of workers nearby to defend the superorganism from outside invaders. The physical contact of foragers with each other in the bee dance results in the recruitment of individuals to food sources.

The breakdown of signaling systems leads to disease. The breakdown of signaling systems between cells of the pancreas that make insulin and target cells throughout the body involved in metabolism leads to type II diabetes. The breakdown in inhibitory signals from queens and larvae leads to the differentiation of a worker type that isn't normally expressed under queenright and broodright conditions, the laying worker. The laying worker expresses a unique physiology and repertoire of behavioral traits that are antisocial and lead to a systemic social disease, the breakdown of order and death of the superorganism.

Cancer is a genetic disease. Type II diabetes is not. Type II diabetes comes when target cells of insulin become resistant to insulin after exposure to elevated levels for long periods of time, a result of our poor diets, and don't respond appropriately to the insulin signal. The pancreas responds to signals to increase insulin production but gets tired and can't keep up. The signaling system breaks down. Invasion of colonies of *Apis mellifera scutellata* by workers of *Apis mellifera capensis* causes a genetic disease. *Capensis* workers no longer respond to the normal signals that maintain social order, a result of changes in the genomes of Cape workers as a consequence of selection in the fynbos. They replicate uncontrolled, consume resources, crowd out other normally functioning bees, and contribute nothing to the well-being or health of the host colony. The superorganism dies—multiple myeloma cancer.

6.5 Parting View

This chapter offers a description of the organizational complexities of honey bee societies. At the end of Chapter 5, I presented the honey bee as a Wheeler superorganism. It's a collection of analogies hung upon the structure of Wheeler's metaphor. It's only as good and as complete as my knowledge of bee behavior and human physiology and my imagination. It's still just a metaphor and does not capture the wonderful kaleidoscopic array of social behavior exhibited by honey bees (see Table 6.1). It also fails to show the mechanisms operating to transform an aggregation of thousands of individual bees in a social entity resembling an organism, the subject of the next chapter.

7

How to Make a Superorganism

> *The [superorganism] concept offers no techniques, measurements, or even definitions by which the intricate phenomena in genetics, behavior, and physiology can be unraveled.*
>
> Edward O. Wilson, *The Insect Societies* (1971, pp. 318–319)

As the epigram attests, E. O. Wilson declared the superorganism concept of no use in 1971 in his landmark book *The Social Insects*. The superorganism was dead. However, in 2008 Bert Hölldobler and E. O. Wilson published *The Superorganism: The Beauty, Elegance, and Strangeness of Insect Societies*, in which Wilson reversed his opinion. Hölldobler is another in the succession of great ant biologists from Harvard University and a longtime close friend and colleague of Wilson's. It's an elegant book full of wonderful stories of ant natural history. They state that the purpose of their book is to revive the superorganism concept with an emphasis on colony-level adaptive traits such as division of labor and communication, perhaps 1,000-meter (m) views. They define a superorganism as follows:

A society such as a eusocial insect colony that possesses features of organization analogous to the physiological properties of single organisms. The eusocial colony is divided into reproductive castes (analogous to gonads) and worker castes (analogous to somatic tissues). . . . Among the thousands of known social insect species, we can find almost every conceivable grade in the division of labor, from little more than competition among nestmates for reproductive status to highly complex systems of specialized subcastes. The level of gradient at which the colony can be called a superorganism is subjective; it may be at the origin of eusociality (preferred by E.O. Wilson), or at a higher level, beyond the "point of no return," in which within-colony competition for reproductive status is greatly reduced or absent (preferred by Bert Hölldobler; see Hölldobler, B., & Wilson, E. O. (2008). *The superorganism: The beauty, elegance, and strangeness of insect societies.* New York, New York: W.W. Norton and Company, p. 514.)

Their definition recognizes varying grades of sociality but still reserves the status of superorganisms for only those approaching the pinnacle of sociality. It relies on organism analogies, as did Wheeler, especially implementation of Weismann's concept of separation of germ line and soma. The "point of no return" referred to in their definition is a point during the evolution of sociality that the workers are so specialized that they're unmated for life, and individual selection no longer acts on them. At that point, colonies within the species can't evolve to states that are less social, the workers have no reproductive options, and their social contracts can't be revoked. They're the body cells, the soma, of the organism.

Superorganisms are built by group (colony) selection. With colony-level selection, colonies survive and reproduce through the production of queens and males and, in some cases, reproductive swarms. Organizational properties of colonies have heritable variation that spans levels of biological organization from genes to the social interactions of colony-mates and emerges as differences in individual and colony phenotypes—from different anatomical castes to communication systems.

A 10,000-m, holistic view of an insect society can be useful for describing the basic organizational structure and activities of insect colonies and may lead to an understanding of how they operate and why they're structured as they are as well as generate new hypotheses of group behavior. Comparative studies of 10,000-m views between species or within species in different environments can lead to some understanding of the evolutionary trajectories and why they're organized as they are, but to deeply understand how they came about and how they function mechanistically and evolve requires a closer look. In the next sections, I paint a closer view of how insect societies are assembled and organized, operate, and evolved. In the following chapter,

I present a 30-year odyssey of discovery, mapping the results of a colony-level selection experiment designed to reveal how colony-level selection affects superorganismal traits and how they evolve.

7.1 How to Make a Worker

Let's go back down to a closer view, return to the birth of an individual worker bee, and look at some of the mechanisms that fated her to her worker role and that regulate her behavior. She emerged into a complex milieu of sensory inputs (stimuli) provided by the social network of the colony, her family. She'll be a reflection of her perception of the information she extracts from the environment and her programed responses based partly on genetic and developmental programing and partly on her own experiences. The template for her behavior (the rules determining her responses to the stimuli in her environment) is determined in large degree by the social contract signed for her by her mother and by her sister nurse bees. A conspiracy against her began when her mother laid the egg that gave rise to her. The egg was laid in a worker brood cell rather than a queen cell, signaling the nurse bees that she was to be raised to be a slavish worker, fated to her predestined role in society. Her sisters were co-conspirators because they share control of a developmental pathway for her that greatly restricts her reproductive options and limits them to produce only males by parthenogenesis. But the conspiracy doesn't end there. Soon after she emerges as an adult, her mother and her larval sisters further manipulate her behavior through her diminished ovaries and place her in behavioral roles that contribute further to the colony division of labor.

Her fated trajectory began as soon as she hatched from the egg. Her sisters began feeding her royal jelly, the same protein substance they feed to her more fortunate sisters that become queens; only her food contained less sugar. The decrease in sugar resulted in lower levels of a specific hormone, juvenile hormone (JH), circulating in her blood. This hormone is needed later in larval development, around the fifth day, to cause changes in the developing tissues that result in a queen instead of a worker. But there are two objectives for the nurse bees: (1) keep the developing larva from becoming a queen and (2) reduce the size of her ovaries, which also reduces the potential for reproductive competition between her, the queen, and other potential egg-laying workers and provides a mechanism through which her behavior as an adult can be manipulated. Sisters with larger ovaries better resist manipulation. Restricting the size of the ovaries, the number of ovariole tubules that make eggs, also requires restricting the total amount of food she receives. Restricted food reduces total body size and ovariole number. So, when the larva is about 2.5 to 4 days old, the nurses restrict the amount of food fed. No longer did she get all the food she could eat, like her queen-destined sisters. She received less food and less sugar. In order to make the final larval molt and become a pupa, she needed more sugar to make more JH. To meet this need, the nurse bees increased the sugar for the last day, still keeping her on a protein-restricted diet; and then, they shut off all food before capping her cell, starving her for her last 2 days of larval development. Her more fortunate queen sisters were fed all the royal jelly they could eat, of elevated sugar concentration, for the whole development time. They were fed a surplus before being capped, and they consumed all they could of the surplus, then pupated. Therefore, it's more difficult to produce a worker than a queen. Queen larvae are fed basically the same diet throughout development, an unlimited amount of royal jelly of constant 12% sugar. Worker larvae are fed a complicated four-stage diet (Table 7.1).

Larvae that are fed more during the fifth larval stage of development have more ovarioles than those that don't. Queens that get unlimited food with more sugar over the whole larval development will have around 300 ovarioles. Workers typically have eight or fewer; however, when they're 3- to 4-day-old larvae, queen and worker larvae have the same number of

Table 7.1 Feeding Programs of Queens and Workers

Age (days)	Stage	Queen diet	Worker diet
0–2.5	L1–L3	High sugar, fed ad lib	Low sugar, fed ad lib
2.5–4	L3–L4	High sugar, fed ad lib	Low sugar, restricted food
4–6	L5	High sugar, fed ad lib	Higher sugar, restricted food
6–7	Prepupa	High sugar, cell capped with surplus of food	Starved, cell capped without food

Note. Workers control development of queens and workers by adjusting the amount of food they can eat and the amount of sugar they put in the food. Queens diets are characterized by an unlimited amount of food (ad lib) available all of the time with a higher sugar concentration. Workers have a four-stage diet.

undeveloped ovarioles, a full queen complement. During the late fifth instar and early pupal development, worker larvae ovarioles undergo programed cell death; they reduce by 97%, while the ovarioles in queen larvae are rescued from cell death by elevated levels of JH. The worker, when she emerges, will have the complement of ovarioles that will influence her behavior for life.

7.2 Worker Ovaries and Behavior

Soon after she emerges she initiates patrolling behavior, foraging for work close to her nursery. She sticks her head in cells containing her undeveloped sisters, larvae that take advantage that she's a slave to her ovaries and begin a process of further manipulation using her reproductive signaling network that operates in concert with her ovaries. Her bondage is complete.

When the worker emerges from her cell, her ovaries normally are in a quiescent state; they aren't activated. But, like many insects, after a few days signals in the form of hormones pass from the ovary to the fat body, preparing the fat body and the ovaries to do their jobs of making proteins and eggs. The fat body consists of cells that cling to the inner surface of the abdomen, like a sheet of cells, with some floating in the blood, hemolymph. They're similar in function to our livers and produce many different kinds of proteins including those needed for egg production, the main one being vitellogenin (VG).

Female insects change their behavior with changes in ovary development. These are developmental relationships between reproductive status and behavior that were inherited from ancient solitary insect ancestors and modified by colony-level selection to function in worker division of labor—a view to the past. For example, female mosquitoes emerge as adults with inactive ovaries. Initially, they feed on nectar, a source of sugar, a carbohydrate needed for physiological maintenance and for energy when seeking hosts. Their ovaries receive signals and become activated along with the fat body. They need protein to produce eggs, so their behavior changes with ovary activation; and they switch from nectar to blood protein foraging, switch from looking for flowers to looking for us. They fly at dawn and dusk, seeking the carbon dioxide from our breath and our body heat. Once they feed on us, they're replete, stomach full, and their behavior changes. They seek out and find a cool, dark place, like the wall behind the toilet in our bathroom, and sit on the vertical surface while their digestive system processes the blood meal and releases our blood proteins into their hemolymph to be used to make egg proteins in the fat body. The egg proteins are released to the hemolymph, taken up by the ovaries to make eggs, and the ovaries swell. Once the ovaries are full of eggs, mosquitoes' behavior changes again, and they seek out water vapor on the surface of a pool of water where they can lay their eggs. Once they have voided themselves of eggs, ovaries empty, they once again seek out flowers for carbohydrate and start the cycle over again.

Worker honey bees also forage for carbohydrate and protein. They often bias their foraging for one or the other, becoming what some call "specialists." Before initiating foraging, they perform different tasks within the nest, tasks that change as they age. This is part of their division of labor. They transition to foraging usually around their third week of life, depending on the race of bee and the hive and external environments. Their age at onset of foraging behavior, and their protein and carbohydrate bias, is related to the state of their ovaries, as we find in the mosquitoes. Workers that are born with larger ovaries, more ovarioles, activate them earlier in life, forage earlier in life (they go through the within nest behavioral transitions faster), and tend to bias their foraging behavior for more protein (pollen), traits controlled to some extent by the queen and larvae. Workers with more ovarioles produce VG at a younger age and in larger amounts that circulate in the hemolymph. After 10 days, the titers of VG decrease rapidly in preparation of foraging. Foragers are stripped of most of their VG titer before beginning. VG is a valuable resource for the colony, and foragers assume very risky jobs with short life expectancies. It's better to pass your surplus body protein to nestmates that can use it than to die with it while foraging.

VG is also used for larval food. Queens produce large quantities of VG that are imported into their ovaries and converted into egg yolk protein used in making eggs. Special proteins, VG receptors, sit on the surface of the ovary that bind to molecules of VG in the blood and transport them through the cell membranes and into the ovary. Workers also produce a large amount of VG, but most of it isn't imported into the ovary but instead is transported by the VG receptors into the hypopharyngeal glands (HPGs), paired glands in the head, where it's converted into royal jelly proteins. The royal jelly is instead fed directly to developing sister larvae, an interesting reassignment of a reproductive function of a major protein, a signature of colony-level selection, a superorganismal trait. The transfer of VG from nurse bee to larva constitutes a reproductive investment of the workers in their mother by raising their sisters; VG is still being used as a reproductive protein. So, existing behavioral and physiological traits were used to build links between reproduction and feeding larvae and define the nurse bee stage of division of labor. The physiological links between ovary and HPG activation are part of reproductive signaling.

The ovaries of adults are controlled by chemical signals from the queen and the larvae. Larvae and the queen produce pheromones that suppress ovary development in workers, controlling against reproductive competition. Young larvae are more effective at suppressing ovary activation than the queen or older larvae. Old and young larvae produce different chemical signals. The pheromone of old larvae is a complex blend of hydrocarbon compounds, while the pheromone of young larvae is dominated by a compound called (E)-β-ocimene. In laboratory tests, (E)-β-ocimene can be substituted for young larvae and produce nearly the same effects as young larvae, while queens can be substituted with the queen mandibular pheromone 9-oxo-2-decenoic acid, with similar effects to a queen. They also activate the development of the HPG, thereby affecting the development of nurse bees and contributing to the worker division of labor. Bees with larger ovaries have higher levels of HPG activation, showing the connection of these two components of reproduction. Bees with larger ovaries, more ovarioles, are also more resistant to suppression of ovary development. In the presence of a queen, they're more often found with some developing eggs. In the absence of a queen, they develop faster into egg layers. Queen and larval effects on HPG, like ovary suppression, aren't equal. Larvae have a stronger effect on ovary suppression and HPG development than queens; young larvae have a greater effect than older larvae.

Older larvae deplete the VG of nurses more rapidly than young larvae because they eat much, much more. By depleting VG, they act on the fat body to shut down the production of VG and increase JH production in the corpora allata, another set of paired glands in the head. VG and JH coregulate each other: VG inhibits the production of JH, and JH inhibits the production of VG. As VG titers decrease in the blood, JH titers rise. Rising JH then shuts down VG production in the fat body, and the bee begins foraging, a part of the mechanism of division of labor that results in the foraging stage that's intimately linked to the nurse stage by common reproductive signaling networks and behavior. As a consequence of these regulatory pathways, old larvae have a strong effect on regulation of the division of labor between nest bees and foragers by shutting down the transcription of VG genes in the fat body, reducing the amount of circulating VG, increasing JH, and initiating foraging behavior in adult bees. Shifting between nursing and foraging behavior represents a switching of reproductive state.

Larval pheromones "release" pollen-foraging behavior in foragers. If you add a frame of larvae to a colony, you can observe at the entrance an increase in the collection of pollen. This increase takes place within minutes. Larvae affect foraging decisions, resulting in a division of labor between pollen and nectar specialization, another level of division of labor. Also, young larvae have a greater effect on foraging division of labor, collecting pollen and nectar, than older larvae. (Note: The worker has now been manipulated at four levels: queen versus worker, nurse bee role, transition to forager, and pollen versus nectar specialist.) The presence of stored pollen inhibits pollen foraging. If you add a frame of pollen to a colony, you can observe a decrease in pollen collecting at the entrance. The two stimuli, larval pheromones (positive) and stored pollen (negative), regulate the decisions of foragers to collect pollen and nectar. A balance between the levels of the larval and stored pollen stimuli and the sensitivity of the foragers to the stimuli regulates the amount of pollen stored in combs in colonies.

Young larvae provide a stronger pollen-foraging stimulus than do older larvae, a consequence of (E)-β-ocimene. (E)-β-ocimene also increases the overall foraging activity of a colony. Mated, egg-laying queens also produce a small amount of (E)-β-ocimene, perhaps partly responsible for the so-called queen morale effect in colonies. Colonies without queens tend to be "lethargic" with reduced foraging activity, while those that are queenright display greater "vigor." The same is noted for broodright colonies versus those that are broodless.

Larvae also "prime" young workers to be pollen foragers 2 to 3 weeks later in life. A primer affects the physiology of the animal in a way that influences future behavior. A "releaser" serves as a stimulus for the immediate expression of a behavior. The mechanism for priming workers to collect pollen is unknown, but exposure of nurse bees to larvae, and especially young larvae and (E)-β-ocimene, increases the likelihood that they'll bias their foraging to pollen collecting 3 weeks later. Bees with more ovarioles are more sensitive to the priming effects of larvae, as they are to the releaser effects.

In summary, bees with more ovarioles produce more VG, have higher levels of ovary activation when young, are more resistant to suppression of ovary activation by the queen and larvae, and have higher stages of HPG development. This results in a positive correlation between ovariole number and activation and HPG development. They're also more sensitive to larval pheromones and the consequences of larval feeding that chaperone them from working at tasks within the nest to foraging outside and that prime them and release pollen-foraging behavior. Nurse bees manipulate the ovary development in larvae; they're at least partially in control of worker behavioral development.

7.3 Worker and Queen Development

BUT DON'T CRY FOR THE WORKER; she's a willing victim. The arms race between larvae and their care providers, the nurses, likely took place over millions of years of evolution as the progenitors of the honey bee evolved their social order from solitary to primitively eusocial to highly eusocial. Worker larvae are fed a complicated four-stage diet, discussed above (section 7.1) The observed pattern is probably a result of an arms race between larvae attempting to circumvent nutritional manipulation by their nursing sisters and their sisters looking for new ways to control their ovary development. In the early stages of honey bee social evolution, the life history was quite different, more like that of many of the other social bees, such as the bumble bees, and vespid wasps, like the pesky yellow jackets that nest in the ground in your yard or in the wall of your house and get into your drinks and eat your food when you dine outdoors. They don't have an anatomically distinct queen caste, except that queens are usually larger and colonies last just 1 year and produce a burst of reproductive males and females at the end of the summer. This kind of reproduction is called the "big-bang strategy." The typical life history is as follows: (1) The colony is initiated in the spring by a solitary queen who was produced at the end of the previous summer. (2) She builds the initial nest alone, defends it, lays eggs, and forages to care for the first batch of larvae, the brood. When the first new bees emerge, they're all females, much smaller than the queen; and they assume the role of a worker. (3) As the colony grows in population over the spring and summer, more food is available for the developing larvae; and they become larger in size. (4) At the end of the season males are also produced along with larger females that mate, the males die, and the females overwinter to become the queens of the next generation. The colony then dwindles and dies before winter.

Workers produced throughout the spring and summer usually don't engage in reproduction. First of all, there are no males; the queen controls their production by fertilizing all of her eggs, a population-wide conspiracy against workers to keep them unmated. Workers can lay unfertilized eggs that become males, and some do later in the season; but their opportunities are limited by physical aggression against them. At some time during the evolution of the honey bee, colony life histories changed from an annual cycle to one that's perennial, persists more than 1 year. With a perennial lifestyle came a new challenge, controlling who reproduces and who doesn't. The old life history just let the natural cycle take care of it; as colonies got larger, more food came in and more was fed to developing larvae, producing larger and larger adults with bigger and bigger ovaries and more reproductive potential. Then, the laying of the unfertilized, male-destined eggs marked the beginning of the reproductive brood.

As a perennial colony, controls were needed on worker reproduction. Worker reproduction reduces the efficiency of the social group by taking individuals out of the workforce and setting up conflict among them for reproductive rights. Reproductive males and females need to be produced at the right time of year for them to disperse, mate, establish their nests and colonies, and be successful. In honey bee workers today, reproductive potential is related to ovary size, as shown in the relationship between ovary size, activation, and egg-laying behavior in laying workers; and ovary size is related to body size: Bees with larger body size have more ovarioles. If you deny food to larvae, the adults are smaller, have smaller ovaries, and are less likely to activate their ovaries and lay eggs (Figure 7.1).

A view to the past suggests an initial state of restricting food to the larvae at critical development times to reduce body size and reproductive capacity. As colonies grew in size, a consequence of living more than one season, queens evolved ovaries with more and more ovarioles to accommodate their egg-laying needs. The normal complement of ovarioles in other bees in the honey bee family Apidae is four per ovary, a total of eight. There are two ways a queen can produce more eggs, either elongate the ovarioles to be a long, factory assembly line or make more, shorter ovarioles and work the production lines in parallel. However, part of the honey bee life history is reproduction by swarming where the old queen flies away. So, unlike most eusocial insects that build large worker populations, a queen must be able to fly even after she has mated and is laying eggs. To be able to fly, queens lose 25% of their body weight rapidly, or the swarm will leave without her. With many ovarioles, she can dump eggs and resorb developing eggs rapidly to prepare for flight.

Body weight and ovariole number are related to each other in the honey bee. The most obvious case is the queen versus workers, with European

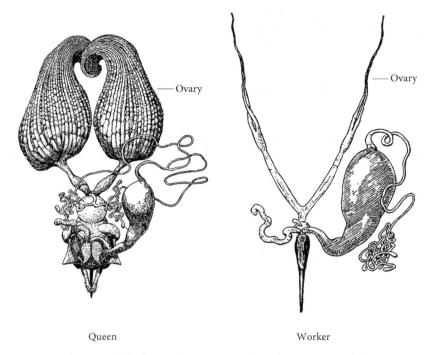

Queen   Worker

Figure 7.1 Ovaries of the honey bee queen and worker. Ovaries of the queen are composed of 150 to 300 individual filaments, the ovarioles. Workers usually have fewer than about 20. Each ovariole is a factory for making eggs. From *Anatomy of the Honey Bee* (pp. 299 and 304; fig. 104 and fig. 106F), by R. E. Snodgrass, 1956, Comstock Publishing Associates. ©1956 with permission from Comstock Publishing Associates, a division of Cornell University Press.

honey bee (EHB) queens weighing nearly twice as much, 175 milligrams (mg) compared to 98 mg, and having on average 22 times more ovarioles, 175 compared to eight. In general, larger body weight results in more ovarioles. However, they aren't completely linked. You can select the two traits independently and get smaller workers with more ovarioles or larger bees with fewer. Africanized honey bee (AHB) populations in Arizona have workers that are smaller in size than the EHBs (90 mg vs. 98 mg), but they also have more ovarioles (13 vs. eight). However, even within AHBs, larger bees have more ovarioles; the same holds for EHBs. The bee size and ovariole number are controlled by different but overlapping developmental processes that are sensitive to feeding at different stages of development.

(Note here that estimates of body weight and ovariole numbers are dependent on the population that's studied, the time of year, and the methods of measurement. Comparisons should be made between groups, queens or workers, AHBs or EHBs, using data collected at the same time, in the same way, for all of the groups being compared. The numbers here for queen ovariole number are significantly lower than those found in bee biology texts but I think are more accurate than most of those that are published because of the way the queens were raised and collected and the ovarioles were counted.)

Honey bee larvae are fed progressively, another step in the evolution of the honey bee's current life history. In bumble bees, eggs are laid on masses of pollen, the larvae hatch, and then consume the pollen. Honey bee nurses consume the pollen and process the pollen proteins in the fat body and HPGs into a viscous secretion that they put in the cells through contact with their mouthparts. Larvae are fed continuously throughout development. Since larvae are unable to access the pollen and depend completely on the nurse bees, they communicate with the nurses with chemical "feed me" signals that provide information about their age, caste, and gender. Nurse bee progenitors of the honey bee adjusted their responses to whatever signals were being used to fit the timing and amount of food delivery to produce the optimally sized worker and restrict her ovary development. Over time, the signals changed as the nurse bees adapted, an arms race between the ever more demanding larvae and manipulative sister nurses changing their nutritional programs in response to changing "feed me" signals.

The arms race between larvae and workers may still be ongoing. When *Apis mellifera capensis* larvae are reared in *Apis mellifera scutellata* colonies, they're able to beg more food from the nurses and end up larger, with more ovarioles. When EHB larvae are raised in an AHB colony, they're larger in size than when raised in their own colonies, even though AHB workers are smaller in size than EHBs. They're able to beg more food. Obviously, the nurse bees and larvae have reached joint solutions to the "problem" of how large they should be and how many ovarioles they should have when under selection in their "home" environments—a genetic truce. Brood signals and feeding responses of nurses evolve together to produce an optimum worker, but relationships are specific to the populations in which they evolved. When you disturb the coevolved relationships between larvae and their nurses, you disrupt normal development and get aberrant bees.

In the absence of a queen, workers select a few lucky larvae to rear as queens, the potential successors to the throne. Larvae compete for recognition from the nurse bees to obtain the food and the feeding program that will make them a queen. The selection of larvae is a consequence, at least to some extent, of their genetic composition. The trait varies within populations and, therefore, is subject to selection. The reproductive advantages to being selected are huge and obvious: inheritance of the nest and all reproductive rights, traits that should be under intense competitive selection. Nurse bees should likewise be under selection to resist manipulation by the larvae and only rear as many queens as is optimal, usually fewer than 20, and to select larvae of the correct age to provide time for development of full queen traits. Nurses determine the ages and nutritional needs of larvae by the chemical signals they send, perhaps an opportunity for larvae to lie and be selected. The interests of the larvae and nurse bees are different, setting up potential conflict between them.

Ultimately natural selection is on the finished product, workers and queens that are best suited for their roles in the social organization. In the case of worker body and ovary size, selection favors workers that have the appropriate behavior and body plans for their roles in the division of labor. A truce was reached between larval workers and nurses, and due to selection on individuals and colonies, workers became willing victims. Colonies with too much conflict between larvae and nurses or that produced workers that had greater capacity to reproduce resulting in competition with the queen and each other, and contributed less or nothing to the division of labor, produced fewer reproductive males and females compared to those that reached a truce and carried the genetic information for cooperation. One of the fascinating aspects of honey bee evolutionary genetics is that the genes that govern worker anatomy, physiology, and behavior may be different from those of the queen, yet they both have the same sets of genes given to them by their parents. The caste-specific expression of the genes allows selection on whole colonies that results in separate selection on queens and workers based on their individual contributions to the colony phenotype. All of the genes responsible for the colony phenotype reside in the queen and her male mates; they're the ultimate targets of natural selection.

7.4 How Selection on Colonies Builds Superorganism Traits

Let's return to the metaphor cartoon of the lemmings in section 4.3 4. Imagine that the lemmings that jump off the cliff are a subpopulation of a larger population of lemmings. It's critical that a certain percentage are pruned every generation to keep from overexploiting resources. Populations are closed; they breed among themselves and are isolated from other populations. If there are genes in a population to cheat, like the lemming with the parachute, it will initially increase in frequency, a consequence of selection on individuals for their individual reproductive success. But as cheaters increase in frequency in the population, the chances of the population surviving decrease. Too many cheaters will not reduce the population sufficiently when it's overcrowded so that it will survive. The entire population will go extinct. Other populations with fewer genes for cheating will survive and outcompete the others.

Another example is our own body. When we develop as embryos, cells divide, each giving rise to a descendent clonal population. But some must die through a process of programed cell death, elective suicide, to reduce the overpopulation of certain types of cells in certain tissues and to sculpt the body plan of the developing embryo. Uncooperative cell lineages can result in deformations, tumors, and system failures that will kill us as higher-level organisms composed of trillions of cooperative cells. Selection at the level of the organism suppresses conflict and competition of cell lineages within. Organisms that suppress the competition successfully reproduce; others die, along with their competing clones, and don't replicate. As a result, we have cellular mechanisms to decrease and eliminate cell line competition. These are examples of two different levels of selection.

The nurses and the larvae jointly share in the developmental program that produces workers and queens, a truce solution, a consequence of selection on individual colonies rather than individual bees. The intricate orchestration of nutritional inputs based on quantity of food and sugar and the timing of food delivery builds an irreversible worker trajectory from an egg in about 6 days.

Students, postdoctoral researchers, and technicians in my lab have been hand raising bees in an incubator for years, using diets derived from commercially available royal jelly. Royal jelly can be produced by the kilogram using colonies that are in hives set up to stimulate queen rearing. Typically two boxes, hive bodies, are used. The bottom one contains the queen, brood, pollen, and honey. A queen excluder separates the two bodies. The queen excluder is either a metal or plastic screen with openings large enough for the smaller workers to pass through but smaller than the thorax of the queen. Very young larvae are grafted from worker cells into queen cell cups that are arranged on wooden bars that fit into specially designed frames. Grafting larvae is a special skill, and it's interesting to watch people who are very proficient. The bars of grafted larvae are placed in the box above the excluder along with frames of pollen and honey. Nurse bees sense the separation of the queen cups from the queen; the excluder disrupts the distribution of the queen's pheromones that inhibit queen cell construction. The nurse bees move up into the top box and raise the larvae to be queens (Figure 7.2). After 3 days, cells are full of royal jelly and are harvested. The bars are removed, the larvae are thrown out, and the royal jelly is removed from the cells.

Most royal jelly sold worldwide comes from China. They have developed stocks of bees that will accept large numbers of queen cells on bars and provide prodigious amounts of food to the cells. In addition, they've developed specific apicultural practices used throughout China by beekeepers that produce royal jelly as a bee product for commercial sale, one that fits the biology of their special bees. For more than 35 years bees have been selected for production of royal jelly using their apicultural practices. Colonies with the largest yields were used as sources of new queens and males in a breeding program that was initiated by Professor Shenglu Chen at Hangzhou University. Professor Chen set up isolated breeding stations on islands in the spectacularly beautiful Lake of a Thousand Islands south of Hangzhou so that he could control the mating of the queens. His colony-level selection program produced queens that were then distributed to beekeepers throughout

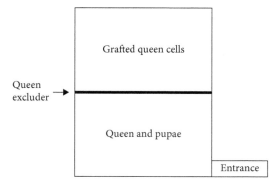

Figure 7.2 Side-view diagram of a commercial cell-builder hive used to raise queens. The two boxes, hive bodies, are separated by a queen excluder to keep the queen confined below. Young larvae are usually removed from the bottom box to stimulate the nurse bees to move up to the upper box where the grafted queen cells are placed.

China to be used in royal jelly production. I have visited Professor Chen and beekeepers in China who use the high royal jelly bees and found the amount of royal jelly harvested to be nearly unbelievable and very fascinating.

Worker larvae and queen larvae are both fed royal jelly, the same stuff. We purchased the Chinese royal jelly and used it for raising larvae in the lab. We prepared a diet that could be fed to larvae that we grafted into petri dishes and sample plates of different kinds. We assumed the role of nurse for raising the larvae to adulthood. We attempted to raise perfect workers and queens paying close attention to the diet composition, frequency of feeding, and changes in the diet as the larvae aged. But we also tried different, lazier, methods of feeding, varying the sugar content, moisture, and protein added to the royal jelly and the frequency at which we fed them. At the same time, we put the combs from which we grafted the larvae back into their hives to be raised by worker nurses, grafted larvae into queen cells, and returned them to a cell-builder colony to raise the queens. When the larvae matured into adults, both those we hand-reared and those raised by nurse bees, we measured several characteristics that are associated with being queens and workers including body weight and number of ovarioles, the most important indicators of reproductive status. We found that an almost continuous range of characters existed in the samples, not just bees that could be called workers or queens. We were very poor nurses; we had a relatively small percent of "perfect" workers and many intermediates. We were better at making queens. Making a good worker is hard work.

The common story in honey bee biology is that queens get fed a different diet from workers, containing some magical, still unknown, substance that "throws a switch" in development that turns on the genes that make a queen; otherwise, they would be workers. However, there's no magic substance, other than sugar, and no master switch. The different traits that distinguish workers from queens, such as the barbed sting, wax glands, pollen baskets, mandibles without notches, smaller body size, smaller spermatheca, and fewer ovarioles, are highly variable, are somewhat under individual control, and differentiate at different times in development in response to nutritional inputs. The master switch is actually the control system of the nurses that care for the larvae and the real-time developmental responses of the worker larvae. Without the special nutritional programs provided to worker larvae, they would become queens by default. Raising a queen is relatively easy: Feed them the same stuff the whole time, and give them all they can eat. In fact, we should be asking the question differently: What is special about how workers are produced? They're the ones that have the most differentiated traits and the most complicated nutritional needs. Variation in all of those traits can be observed in the cloud of bee phenotypes we produced by being bad nurses, delivering the wrong food at the wrong times (Figure 7.3).

Members of my laboratory reared literally thousands of bees in the lab from just-hatched larvae. All larvae of a given treatment were fed the same diet; food was identical, from the same sources. Diets were varied from treatment to treatment with respect to the amounts of components of the diets and the total amount fed, but all larvae were fed the same "stuff." The dots in

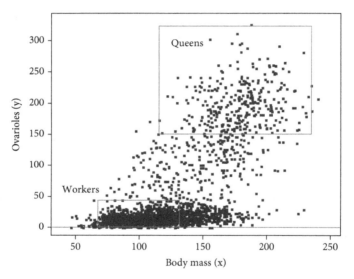

Figure 7.3 Relationship between body mass and number of ovarioles for queen and worker honey bees. Body mass and ovariole number are considered two of the most important differences between the phenotypes of queens and workers. Ovariole number is significantly correlated with body size, but the relationship is loose, with lots of variability. Differences between queens and workers are determined by the food fed to them by worker nurse bees; either can be raised from any egg. For decades, bee researchers believed that differences in development between queens and workers were due to some kind of special substance fed to queens but not workers that turned on a switch leading to a very different developmental program that resulted in a queen. "Development and Evolution of Caste Dimorphism in Honey Bees—A Modeling Approach," by O. Leimar et al., 2012, *Ecology and Evolution, 2*, 3098–3109, fig. 5. © 2012. Wiley Online Library (https://onlinelibrary.wiley.com/doi/10.1002/ece3.414).

Figure 7.3 represent the body mass and number of ovarioles for single individuals raised. You can see that there are clearly more than two phenotypes, worker and queen. In fact, there's a continuous cloud of phenotypes. The two boxes capture the phenotypic space of workers and queens that are raised in a hive by worker bees. The boxes represent the full range of phenotypes of thousands of bees we weighed and dissected. Clearly, there are only two types when the nurse bees are in control, compared to the cloud produced when raised by us in the lab. The figure demonstrates that distinct queen and worker phenotypes are controlled by the feeding behavior of the nurse bees, not by a special substance.

7.5 How Complex Social Behavior Evolves

A worker walks on the vertical surface of the comb within the brood nest, close to where she emerged as an adult, sampling the stimuli that surround her, some releasing behavioral responses. She has no set of instructions about what to do; each encounter with a stimulus is a different response or no response. She's born into the middle of the nest and undergoes a centripetal movement while she ages, foraging for work in different locations, progressively distant from the nest center, the brood nest. As she changes location, the stimuli to which she responds, and her thresholds of response to those stimuli, change, establishing her changing roles in the age-related division of labor among her sisters.

Temporal Changes in Behavior

When viewed by an observer, it appears that the bees undergo a temporal age-based division of labor, changing their tasks and the locations in the nest where they performed them. This is called *age* or *temporal polyethism*. In the late 1980s to 1990s, there were challenges to the model of age-related division of labor. The challenges were that there really isn't such a thing as age division of labor that depends on internal changes of workers, no physiological drivers, no age-paced clock, a very existential view of workers. Age-related division of labor is an epiphenomenon of social self-organization. Worker changes in behavioral roles as they age are consequences of them foraging for work. The idea was that adult workers are "born" into the center of the nest.

When they emerge, they begin foraging for work where they are. Patrolling behavior, walking around on the combs, is the mechanism they use to encounter behavior-releasing stimuli. Patrolling is the major activity of bees. So, since they just came out of a cell that's surrounded by lots of other cells, also with new adults emerging, they begin by cleaning the nursery. But as more and more come out, they're pushed out of that part of the brood nest and start looking for other work close by. There they encounter larvae and begin feeding and caring for them. They remain in that part of the nest feeding and caring for larvae until overrun by younger bees moving in, forcing them out of the brood nest and into the food storage areas, etc., until they're forced out of the nest to forage. Foragers die and create a vacuum to be filled by younger bees as they're pushed out of the nest periphery, stimulating them to initiate foraging. So, births in the middle force the tide of maturing workers outward to fill the void produced by dying foragers. Although an interesting idea, and one that could explain a complex social structure with a simple self-organizing, automatous process, a popular modeling strategy of the time for ant colonies, it doesn't really work for bees.

It's easy to observe if you mark newly emerged bees and place them in an observation hive. An observation hive is a special kind of hive built for conducting behavioral research or, sometimes, by hobby beekeepers who want to watch their bees in the privacy of their study or other indoor venue. It's constructed so that basic commercial frames with combs are stacked up vertically, one on top of the other, usually two to four, with glass sides allowing observation of the internal workings of the colony. They're also equipped with an entrance to the outside of the house or other structure so that the bees can leave and forage. If you mark newly emerged bees with paint, place them on top of the top frame, and wait, the young workers will all find their way to comb that contains the larvae and stay there. They didn't forage for work where they were; they moved around until they found the correct location in the nest corresponding to their stage of age development.

Response Thresholds and Social Organization

Division of labor is an epiphenomenon of self-organization. It's simply not possible to have an overarching control system to control tens of thousands of individuals, as was proposed for termites by Marais, emanating from the "brain" queen. Changes in location associated with changes in age bring our worker into contact with different stimuli. As she patrols, she's bombarded with chemical and tactile stimuli. Visual stimuli are limited inside the nest due to the nest location within a sealed cavity. However, we can't rule out visual stimuli completely because light can still enter the nest from the entrance and cracks in the cavity. The perception of the stimuli begins at the different sensory organs located on different parts of her body: tactile (touch), olfactory (smell), gustatory (taste), and visual. Olfactory receptors are located on the antennae; gustatory are located on the antennae, mouthparts, and feet; and touch receptors are concentrated on the antennae but found over the whole body. The specific stimulus, say the larval pheromone (E)-β-ocimene, causes the receptor to produce an electrical signal that's sent down the antenna and into the brain, where it's processed. The electrical signals are related to the strength of the stimulus. If the stimulus strength is strong enough, then the brain sends back a signal that triggers a motor response, a prepackaged stereotypic set of neural electrical signals that travel from the central nervous system to motor neurons (nerves that stimulate muscles) and result in a stereotypic behavioral response to the stimulus; if not, she ignores it. The minimum signal strength that releases a behavioral response is the bee's response threshold to that stimulus.

It's easy to demonstrate what a response threshold looks like by using the proboscis extension reflex (PER) assay (Figure 1.4). Restrain a worker in a small tube so that she can't move her body but her head extends from the tube, giving her free movement of her antennae and proboscis. Then, touch droplets of sugar water to her antennae. We use a serial dilution and present it in increasing concentrations, starting with pure water. Initially, the worker doesn't respond; but when offered a concentration that's high enough, she responds by reflexively extending her proboscis, her tongue. It's the end result of the sugar receptors on the antennae sending electrical signals to the brain where they're processed and motor response signals being sent back to the muscles that control the proboscis. The response is stereotypic, meaning always the same, a result of prepackaged electrical impulses that originate in the central nervous system. They travel down pathways of nerves to the muscles where the motor neurons cause the muscles to contract, resulting in movement—behavior. The same package of electrical signals will be generated time and time again, producing the same behavior over and over again. One can record the lowest concentration to which they respond, and that's their response threshold.

The response threshold to sugar has proven to be a useful indicator of the differential tuning of the central nervous system to stimulus inputs from the environment. Each bee has a profile of response thresholds to different behavior, releasing and priming stimuli. The profile changes with age and experience. As bees age and come into contact with new stimuli in different parts of the nest, their response probabilities change. For example, young bees less than about 5 days old are unlikely to respond to alarm pheromone and defend the nest. Bees that are of foraging age don't respond to larval pheromones by feeding larvae. Response thresholds are tuned for location-specific and task-specific stimuli. Sugar response thresholds change with age, exposure to pheromones, and levels of hormones in the blood that are associated with age-based division of labor and differ among individuals performing different tasks. It's one of a multitude that modulate the behavior of bees and shows us how they're related to changes in behavioral roles. Pollen foragers have lower response thresholds to sugar than do nectar foragers and are more responsive to pollen when offered in a PER test. Bees that return empty from a foraging trip are the least responsive to sugar and have the highest thresholds, perhaps related to responses to other stimuli that set them in roles as scout bees.

Sugar response thresholds also differ among workers on the basis of the size and state of their ovaries, showing a link between reproductive state and behavior. Workers with more ovarioles are more sensitive to sugar when they're young than are bees with fewer, suggesting different behavioral "tuning" based on ovary state. Behavioral links to ovary states are obvious when you look at queenless bees and see the different roles assumed by bees that are based on ovary development such as egg eaters, egg layers, and nurses. In addition, queens change their behavior once they have mated and their ovaries are making eggs. Virgin queens are very active, often anthropomorphically called "nervous," until they make their mating flight. Afterward, they settle down into an egg-laying behavioral routine.

Division of labor self-organizes and occurs when the response of a bee with a low threshold reduces the level of stimulus that others might encounter. This is a negative correlation between the response and the stimulus. I'll begin with a non-bee example. Imagine two people living together in an apartment, sharing the apartment stimulus environment. One of them has a low response threshold for dirty dishes; the other's is higher. For one, only a few dirty dishes in the sink stimulates him to wash them. The sight of the dirty dishes is the stimulus. When the dishes are washed, the stimulus level is decreased, thereby "walling off" the other with a higher response threshold from doing that task. The stimulus never gets high enough for her to respond.

Another example is thermoregulation in a bee's nest. Different bees have different response thresholds to temperature, releasing fanning behavior. Bees fan in response to heat and circulate the air, thus cooling the nest interior. When those bees with lower thresholds respond, they decrease the temperature, "walling off" others with higher thresholds from performing that task. They're free to engage in other tasks for which they're more responsive. A division of labor is formed.

A foraging division of labor may also emerge when a potential forager patrols over a comb that contains young larvae and some pollen on the edge of the brood nest. While patrolling, she samples the brood pheromone stimuli that can release pollen foraging and the stored pollen stimuli that can inhibit it. She integrates the inputs into her brain from her sensory receptors. If the signals from the larval pheromones are strong enough and the inhibiting signals from the stored pollen weak enough, her sensory-response system will be tuned and she'll get motor response signals passed down that release pollen-foraging behavior. She exits the nest, collects pollen, and returns, storing the pollen she collected in the combs near the larvae. By storing the pollen, she increases the inhibition cues provided by the stored pollen and potentially affects the sensory inputs to other potential foragers, reducing the likelihood they'll forage for pollen. Central control of behavior isn't necessary. A division of labor emerges because bees live together in a nest, get information for themselves from their own local activities, respond to stimuli, and change the stimulus level for others in the nest when they respond. Nothing else is needed. Internal changes in hormonal signals associated with age, exposure to external pheromone signals, and experience modulate response threshold profiles and behavior. The evolution of specific social structures of division of labor is a history of shaping sensory response systems in time and space to fit the needs of the colony in specific environments. Much of the history is captured in the ovaries, who controls them and how, and the reproductive signaling systems of bees.

7.6 Parting View

In this chapter, I provided a general description of some of the fabric of superorganism evolution. I showed how social behavior can emerge from the mechanisms of individual behavior and how those mechanisms can be influenced by social interactions among nestmates—in the case of the honey bee, her sisters. I showed how developmental changes can occur that affect the behavior of individuals, how development is manipulated by nestmates, how those manipulations make them vulnerable to further manipulations as adults, and how opposing levels of selection, colony and individual, can occur with colony-level selection shaping social behavior. In the next chapter, I'll explore how complex social structures evolve by using human-assisted selective breeding as an analogue of natural selection.

8

How a Superorganism Evolves

Nothing in biology makes sense except in the light of evolution.
Theodosius Dobzhansky (1973)

Theodosius Dobzhansky was one of the architects of the modern theory of evolution, linking Charles Darwin's focus on the whole animal phenotype to what was known about genetics at the time. He believed that the only way you could understand and explain the biological world around you is by looking at it as being derived through genetic selection, either natural or human-assisted. He was one of a school of evolutionary biologists using laboratory selection of fruit flies to understand evolutionary processes. What they learned in the laboratory they took to the field to study natural populations, seeking validation of their experimental results. Their selection experiments mapped the anatomical, physiological, and behavioral traits of laboratory and natural populations of fruit flies down to individual genes on individual chromosomes.

Darwin used the results of selection experiments and evidence drawn from human-assisted selection of breeds of animals and varieties of plants to usher strong support for his fledgling theory of evolution by natural selection. He saw human-assisted selection as an analogue of natural selection and a powerful method to test his theory. He studied breeds of pigeons in depth, claiming to have raised and bred every variety he could get his hands on, and documented differences among breeds for beaks, eyelids, nostrils, gape of the mouth, overall size, feather color and pattern, flying behavior, and more. Breeds looked so different from each other they could have been mistaken for different species, all products of selection by the hands of pigeon enthusiasts.

When you look at other domestic breeds of animals today, it confirms Darwin's observations. Today, there are more than 600 distinguishable breeds of cattle, all descendants of the aurochs that roamed the Old World from Siberia to Europe and the subject of many cave drawings of prehistoric humans. Extinct since 1627, it was bred for meat and milk and as a draft animal. Their use for draft is believed to be responsible for the invention of the wheel. The extinct ancestral auroch stock was reconstructed in Germany in the 1920s and 1930s by two brothers, Heinz and Lutz Heck, who were zookeepers in Berlin and Munich. They combined different breeds and selected for the ancestral traits with the intent to reconstruct what they thought was the perfect supercattle. The resulting breed, Heck cattle, are still bred in Europe today.

Chickens provide another excellent example. The ancestor of the domestic fowl is the jungle fowl, which lives in India and Sri Lanka today. I had the privilege to see wild jungle fowl during a recent visit to Sri Lanka; they looked just like tree-climbing chickens to me. Domestication of the jungle fowl occurred in India and China around 3,000 years ago. They were bred primarily for cock fighting and then later for meat and eggs. Most modern breeds of chicken were bred in the last 100 years in Europe and the United States, using standard methods of human-assisted selective breeding. Today, there are around 65 distinct breeds. They vary in many traits such as body size and size of combs and wattles, feather coloration, and meat and egg production.

Drawing on the models provided by Darwin and Dobzhansky, I'll tell you about my selection experiment, a single-minded, career-long passion to understand how complex social behavior evolves. The experiment is a test and validation of a view of how colony-level selection reshapes the social structure of colonies by effecting changes at all levels of biological organization from the gene to the superorganism. I was the force of selection, a selective force analogous to natural selection, selecting the most fit colonies, those that best met my criteria, to survive and reproduce each generation. The story spans 23 years and involves 43 generations of selection for a single trait: the amount of pollen stored in the comb. Here I'll only tell a small part of the story, I have already told the longer version in *The Spirit of the Hive: The Mechanisms of Social Evolution*; another opus is unnecessary.

8.1 A Selection Experiment

Genes reside as parts of chromosomes in cell nuclei deep inside individual cells within tissues within the bodies of the workers, hidden from the direct effects of the environment and natural selection. There are no genes "for" social behavior. There are genes and gene regulatory systems that control the production of proteins that affect anatomy and physiology including sensory-response systems that affect individual behavior and the social interactions of nestmates. For colony structures such as social behavior to evolve, changes in gene expression and gene functions must change properties of individual workers that alter their behavior and form the social phenotype. The social phenotype is what you see as social behavior. It could be the defensive response of a honey bee colony when you kick the hive (not recommended) or, as I discuss in upcoming sections, the quantities of stored food and where they put it. Therefore, selection on colonies (colony-level selection) must act on genes and gene regulation organizationally far from the colony-level phenotype, and genes must act through many levels of biological organization to contribute to colony phenotypes. Selection itself is on the final product, the colony phenotype. Because of the complexity of systems at each level of organization and their interactions, there are probably many different ways to arrive at the same solution. Any one solution, relationships of genes, physiology, and anatomy, reflects the individual, personal history of colonies within a given population. In the case of the honey bee, a population would be a collection of colonies that are able to breed with each other, by producing virgin queens and drones. This is called the *effective breeding population*, and estimates of its size vary from around 500 to 3,000 depending on geographical location or, in the case of honey bee breeding programs using artificial insemination, the size of the pool of parent colonies every generation.

The Colony Phenotype

One difficulty with the superorganism metaphor is its attempt to define a social group as an entity called a superorganism. As Bert Hölldobler told me, "Nature does not fit neatly packed into drawer boxes." That then requires making claims about exactly what traits are necessary and sufficient to move into the "superorganism" classification. Edward O. Wilson believes that all eusocial societies qualify as superorganisms. Hölldobler believes that to be a superorganism, populations must evolve to the state where workers are beyond the point of no return; workers have lost their ability to mate and reproduce. You see an extreme level of caste differentiation in only an elite group of ants, such as the fire ant *Solenopsis invicta*, where the workers lack ovaries. In many species of bees and wasps, workers don't mate and therefore can't produce anything but haploid males by parthenogenetic mechanisms, and then normally only when the queen is absent, perhaps beyond the point of no return. But then, when considering the point of no return in social evolution, we need to consider the Cape honey bee, which shows us that even when we think there's no going backward in sociality, it can happen. They have an alternative life history of being invading parasites, clearly a successful antisocial behavior. It reminds me of the line in the movie *Jurassic Park* (1993) where Malcomb, the complex systems mathematician, said that "life . . . finds a way."

The observation that these two leaders in sociobiology have different opinions of what constitutes a superorganism, explicitly pointed out by them in their book *The Superorganism: The Beauty, Elegance, and Strangeness of Insect Societies*, points out the problem. To avoid it, I think it's best not to try to stick insect societies that represent an amazing span of diversity into individual boxes with restrictive labels. I find it more interesting and rewarding to look for superorganismal traits. In my definition, these are colony traits that were selected by colony-level selection. The trait I studied is a social phenotype, the amount of pollen stored in combs of honey bees.

A phenotype is what you observe and are able to describe and/or measure. It's a result of genes that interact with each other and with the environment. For example, I am about 175 centimeters tall, about average for my family lineage. We share genes in common that affect development associated with stature. But other factors are also important, such as nutrition. As our mothers often warned us, "If you don't eat your vegetables, it will stunt your growth."

Colony-Level Selection

Honey bees store surplus pollen. The amount of pollen stored is regulated by colonies; it's a social trait, a colony-level phenotype, the product of the collective genotypes of colony members and the environment in which they live and interact. It's easier to get a good identifiable and repeatable phenotype if there's some target value the colony "seeks." Honey isn't regulated. Colonies will store honey as long as there's nectar available, and they'll stick it in every available cell. Pollen is stored in a more restricted area of the brood nest. Stored pollen inhibits foragers from collecting it. The chemicals, pheromones, produced by young larvae stimulate pollen foraging. The amount of stored pollen and the location where it's stored are consequences of complex interactions of nestmates that include larvae and adult workers of different ages. In 1989, Kim Fondrk and I initiated a selective breeding program at the University of California Davis, designed to test how colony social behavior is organized, its mechanisms, and to learn more about how it could have evolved. I wanted to study the origin and evolution of superorganismal traits.

For 43 generations spanning 23 years, we selected for one single thing, the amount of pollen stored in combs in colonies of honey bees, colony-level selection on a social phenotypic trait. Pollen is stored close to the brood in a thin shell across the top and down the sides of the brood nest. Honey is stored outside the shell of pollen and will occupy space to the edges of the nest. Each generation, usually one or two per year, we measured the area of comb containing pollen in colonies in our breeding program and selected parent colonies to produce virgin queens and drones that were used to instrumentally inseminate the queens. We established two separate populations, one that was bred for increasing amounts of stored pollen, the high pollen-hoarding strain, and one that was selected for decreasing amounts, the low pollen-hoarding strain. Selection never varied, always for the single colony trait. Each generation we looked at how we'd changed colony structure with regard to the queen, workers, and nest organization. We looked additionally at the anatomy, physiology, development, and behavior of individual workers and mapped genes within their genomes that were responsible for the differences we saw between our selected strains. We mapped the effects of colony-level selection from genes to complex social interactions.

After just five generations of selection, the high-strain bees stored six times more pollen than the low strain, eight times more after 11 generations, and 16 times more by generation 20. This was a change in a colony-level, superorganismal, trait that could only occur as a consequence of selection on colonies. It was a consequence of changing the behavior of the individual workers in the colonies and the interactions among them. Associated with the increase in levels of stored pollen was a shift in the investment of the thousands of foragers toward collecting pollen or nectar. The high-strain colonies had a significantly higher proportion of foragers collecting pollen than did the low strain. The total numbers of foragers did not change between the strains.

Larval Development—A Sociogenome

To achieve changes in foraging behavior, we altered two aspects of what I will call colony physiology. In this case, they have self-organized control systems that regulate development and nutrition. Our selection program altered the size of the ovaries of the workers such that the high-strain workers have more ovarioles. This was accomplished by the substitution of genes and genetic regulatory elements that alter the physiology and behavior of the nurse bees to provide the nutritional inputs to the developing larvae and the developmental responses of the larvae to the nutrition they received. Both changed. They coevolved toward the optimal solution of body mass and ovariole number. We know this because when we raised the high-strain larvae in colonies of the low strain, they had low-strain nurses, were larger in size, and had more ovarioles than when raised by their own nurses. Low-strain larvae when raised by high-strain nurses were smaller in size and had fewer ovarioles than when raised by their own nurses. The effects were dependent on life stage. Total body weight depended most on who raised them during the early stages of larval development (L1–L4), and ovariole number was more dependent on who raised them as older larvae (L5). The orchestration of the timing and quantity of food assisted the developmental growth program of the larvae. The combined genetic control of the nurse bees and larvae during larval development I call the *sociogenome*, a consequence of selection on the colony, a superorganismal trait.

A direct developmental cause of the differences in ovariole number between the high and low strains was their response to the nutritional regime of their nurses during the late larval stages. High-strain larvae, by producing more juvenile hormone that circulated in their blood, rescued more ovarioles from programmed cell death. High-strain workers typically have more ovarioles than low-strain workers, about 12 ovarioles compared to six in one study. The difference in juvenile hormone titers in the late developing larvae is a signature of colony-level selection for stored pollen, the causal trait.

Regulation of Stored Pollen

Colony-level selection had effects deep down inside individual workers, down multiple layers of biological organization, affecting anatomy, physiology, sensory-response systems, behavior, and social interactions involved in the regulation of stored pollen. Our selection program changed response thresholds of workers of the high and low strains for several things we were able to measure including sugar, light, stored pollen, and larval pheromones. Response thresholds to sugar affected foraging decisions associated with loading on individual foraging trips for pollen, nectar, or water. High-strain bees were more responsive to lower concentrations of sugar solution, resulting in them accepting nectar in flowers of a lower concentration than the low-strain bees. They didn't prefer the lower concentrations; they just didn't reject them. They were also more likely to respond to pure water and more likely to collect water when the colonies needed it. Low-strain bees were more discriminating in the nectar they collected and more likely to return from a foraging trip empty; nothing was good enough.

High-strain bees were also more sensitive to the chemical signals, pheromones, produced by larvae. They were more responsive to both the releaser and the primer effects. They foraged at a younger age, perhaps at least partly the result of their greater sensitivity to the brood pheromones that stimulate foraging. They were also less sensitive to the inhibiting effects of stored pollen. The combination of these stimuli they encounter in the nest adjusts their "loading algorithms," the decisions they make when they forage as to how much pollen and nectar to collect. A bee can only carry so much on any given trip. If she loads with only nectar, or water, she can carry about 60 milligrams (mg). If she loads only with pollen, her maximum load is 30 mg. Before the bees leave the nest to forage, they make an assessment of the stimuli, brood and stored pollen, and "set their targets" for pollen and nectar collecting (not consciously, of course) by adjusting their response thresholds for stimuli associated with collecting each. However, while foraging, they get input from the floral sources about the quality of the resource that also influences their final decisions about when to return to the nest. Total loading also depends on the availability and quality of the resource. When the concentration of the nectar is high, they'll collect more nectar and less pollen than when it's low. It's very evident as the day progresses and foragers return with progressively smaller loads of pollen and larger loads of nectar of higher sugar concentration, a consequence of the daily pattern of flowers delivering pollen early, then more nectar as the day warms up. The evaporation of nectar increases sugar concentration (Figure 8.1).

The social demographies of high and low colonies have characteristic differences, consequences of selection. High-strain bees emerge as adults in a higher stage of reproductive activation than bees of the low strain. Many emerge with ovaries that are already activated and ready to take up vitellogenin. They also produce more vitellogenin within a few days of emergence and, when separated from the queen, begin laying eggs sooner than low-strain bees. Their higher activation state results in them advancing through stages of behavioral development faster and with greater sensitivity to brood pheromones, an earlier onset of foraging. As a consequence, they don't live as long as low-strain bees. Honey bees have what's called a *two-stage mortality curve*. Deaths are less common for bees when they work in a nest. Once they begin foraging, the death rate increases. The total life span of a worker is composed of her life as a nest bee followed by her life as a forager. Her age when she begins foraging marks an important transition point in her life and is the main determinant of her life expectancy. If a bee stays in the nest, transitions to foraging later in life, she'll live longer; but her total foraging life will be less because every day she's working in the nest decreases the time she has as a forager by about one third of a day. Bees that forage at earlier ages, therefore, have shorter life spans. Low-strain bees live longer than do the high-strain bees because they initiate foraging later in life (Figure 8.2).

Another social characteristic of high- and low-strain bees resulting from our selection is the difference in scouting and recruitment behavior. Some bees forage for resources without having been recruited. They fly out of the nest and seek new resources. When they return, they perform recruitment dances for the resource they found, giving distance and directional information. High-strain scouts are more likely than low-strain scouts to return with pollen and recruit for new pollen sources. New potential high-strain recruits are more likely to attend the recruitment dances of pollen foragers, thereby biasing their foraging for pollen.

Genes

For evolution to occur and for an effective response to artificial selection, there must be changes in the genes that affect the behavior. Evolution is the substitution of alternative forms of genes and elements that regulate

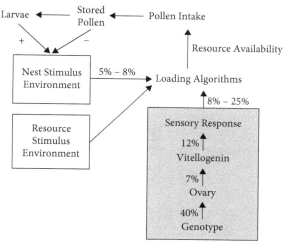

Figure 8.1 Social regulation of stored pollen. Honey bee colonies regulate the amount of pollen stored in the comb. The amount of stored pollen in a nest at any given time is a function of the rate of consumption of pollen by nurse bees, which use the pollen proteins to feed larvae, and the amount of pollen being collected by foragers, the pollen intake. Larval pheromones and stored pollen are major components of the nest pollen-foraging stimulus environment. Workers sample and assess the stimulus environment of the nest and make foraging decisions according to their assessments. Foraging decisions are based on loading algorithms, for example, how much pollen and nectar to collect, that are used on their subsequent foraging trips. Both the nest and resource environments contribute stimuli that determine loading. In addition, loading algorithms are influenced by factors internal to the bee, such as her genotype and reproductive physiology that "tune" her sensory response physiology, shown in the shaded box. Loading algorithms bias foraging behavior, but the availability of resources puts large constraints on the actual loads collected. The percentages shown over the tops of some of the arrows that connect boxes were derived from many studies, over many years, where we assessed the percent of the variation we observed for the variable at the head of the arrow that could be attributed to the variable at the arrow's trail. For instance, nest stimulus environment can explain 5%–8% of differences found in loading algorithms, while sensory responses explained 8%–25%, depending on the experiment. Reprinted from "Regulation of Honeybee Worker (*Apis mellifera*) Life Histories by Vitellogenin," by G. Amdam, D. E. Ihle, and R. E. Page, in D. W. Pfaff, A. P. Arnold, A. M. Eigen, S. E. Fahrbach, and R. T. Rubin (Eds.), *Hormones, Brain, and Behavior* (2nd ed., vol. 2, pp. 1003–1005, fig. 7), 2009, Elsevier. © 2009 with permission from Elsevier (https://www.sciencedirect.com/science/article/pii/B9780080887838000292).

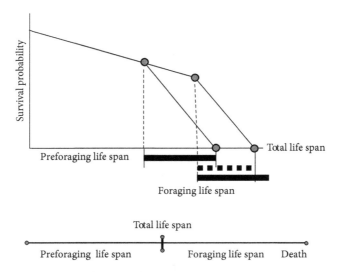

Figure 8.2 Partitioned life span of the worker honey bee. The life span of a worker can be divided into two parts by her age when she initiates foraging. The transition from working inside the nest to foraging can occur as early as 5 days or as late as 40 days. Some bees die before they ever begin. However, most begin foraging around their third week of life. The total life span can be partitioned as shown in the lower line of the figure into the preforaging (within nest) life span and the foraging life span, followed by death. Survival probability in the upper figure is the chance that a bee will be alive on any given day. The slope of the survival probability line changes for worker honey bees once they initiate foraging (shown by the gray dots on the survival line); they have a two-stage survival probability. Bees are dying faster once they initiate foraging, probably due to the additional wear and tear on foragers compared to bees that are tucked snugly in the protective nest. Because of the two-stage survival, bees that transition to foraging later in life have a longer total life span but a shorter foraging life. That can be seen by comparing the thick solid black line below the upper figure with the thick dashed line. The solid line is the foraging life span of the worker that foraged earlier in life, and the dashed line is the foraging life span of the worker that initiated foraging later. The dark solid line below the dashed line is again the foraging life span of the bee that initiated foraging earlier and shows the difference in total foraging life span between them.

them. The populations of high- and low-strain colonies underwent gene substitutions at many places throughout the genome that we were able to map. We identified many possible genes involved in differences between the strains and built hypothetical gene networks that work together to orchestrate the changes at the different levels. Networks were identified that affect each of the levels of biological organization we explored from the social trait, the amount of pollen stored, through sensory-response physiology, reproductive anatomy, hormonal signaling, and larval development. Selection on the colonies effected changes across all levels, as predicted. Genes work together in an environment of interacting proteins that orchestrate development and calibrate the sensory-response systems for behavior. In the past, we had a naive view of the effects of genes that has been likened to "bean bag" genetics, a gene for this trait and a gene for that one. We now know that the actions of the genome, the collection of all the genes and their controlling elements, are far more complex. They form interacting networks that couple with the environment to produce phenotypes that we observe. Many different networks of genes may result in the same phenotypes. If we were to conduct our 23-year selection program all over again—heaven forbid—it's likely that we'd uncover different networks that yield the same changes in the social behavior of the honey bee superorganism.

8.2 Reverse Engineering Society

A few years ago, I was in Shanghai, China, for some official responsibilities as the university provost for Arizona State University (ASU). As the chief academic officer of ASU, I was there to confer professional master's degrees in business administration to Chinese students who had enrolled and completed the requirements of a program we offered there. Since many of the students were officials in the Chinese government or part of the new billionaire entrepreneurial class in China, I was treated well, due to my high administrative position at ASU. I had privileges such as staying at the Presidential Penthouse Suite at the Radisson Hotel and a personal tour of the top of the Shanghai Tower, tallest building in China, second tallest in the world at 632 meters, 128 stories. It has the world's highest outside observation deck. The building wasn't finished, but that didn't stop us from summiting it and standing with barely any protection at the top of Shanghai. I also got to meet with some of the people who plan and oversee the economy.

We went to a special area of Shanghai called a "free enterprise zone" where I met with members of the Chinese Ministry of Finance, the organization that manages macroeconomic policies, basically the economy. China has been a marvel of economic success, with growth averaging close to 10% over the past four decades, greatly improving the living conditions of its citizens. The free enterprise zone, as explained to me, is where the ministry relaxes certain restrictions on trade and imposes others, a testing ground for the economy. They then study the effects of the changes and determine if the changes should be implemented for the whole country. They discussed the success of economic policies that increased production and exports, and population demographic changes that resulted from the changes in the economy, like influx of rural residents to the cities. Then they said now they needed to have more consumers within China to continue to build the economy and couldn't depend only on exports, and I noted that they lifted the one child limit per couple; now they can have three, more consumers being produced. Then, it struck me what they were doing: They were reverse-engineering society by manipulating the economy. That's exactly what I did when I selected for pollen-hoarding behavior in bees. I asked them if that was the case; sheepishly, one of them agreed, though I'm not sure they understood the question.

8.3 Parting View

I selected for one thing for 23 years, the amount of pollen stored in the combs. I reshaped the economy of the hive, the stored honey and pollen. I reverse-engineered society, changing many features of the social structure of the colony, though I didn't select directly for them. In the case of the high strain, the following changes resulted: more pollen foragers, shorter life span, earlier onset of foraging, communication between foragers and potential recruit bees, communication between larvae and nurse bees, organization of the nest, and probably many other things I didn't think about measuring. Reflecting back on my selection experiment, and what I learned in Shanghai, perhaps the evolutionary process and outcomes are not so different between us and bees. I'll reflect more on that in the Epilogue.

9

The Song of the Queen

When the air is wine and the wind is free
And the morning sits on the lovely lea
And sunlight ripples on every tree,
Then love-in-air is the thing for me—
   I'm a bee,
    I'm a ravishing, rollicking, young queen bee,
   That's me.
I wish to state that I think it's great,
Oh, it's simply rare in the upper air,
   It's the place to pair
   With a bee,
Let old geneticists plot and plan,
They're stuffy to a man;
Let gossips whisper behind their fan.
   (Oh, she does?
   Buzz, buzz, buzz!)
My nuptial flight is sheer delight;
I'm a giddy girl who likes to swirl,
   To fly and soar
   And fly some more,
    I'm a bee.
And I wish to state that I'll always mate
   With whatever drone I encounter

        E. B. White, "Song of the Queen Bee" (1945)

9.1 Coming Out

It's two o'clock in the afternoon. A gentle breeze greets the young queen as she sits at the entrance of the nest she inherited just days ago when her mother abandoned her, departing with an entourage of her sisters, a superorganism divided. The day is warm, no need to prewarm the muscles in her thorax, muscles that will probably be used only two or three times in her long life, a life that's short to her, about 1 to 2 years, but long for her shorter-lived sisters who can only hope to live a meager 6 weeks—disposable cells of the superorganism body. She takes flight and begins her search for that special place, that rendezvous point where she'll have sexual encounters with perhaps 20 males, each invited by her aphrodisiac perfume, occurring in rapid succession as she flies through the midst of thousands, scrambling to be the first, or the next, to copulate with her. In the process, each will commit self-mutilation and suicide. This is the nuptial flight of the queen.

As tawdry as it sounds, a hard-core scientific description does little better. Many years ago, in the early 1980s, I was interviewed for a national television news magazine. At the time, I was engaged in a honey bee breeding program with the US Department of Agriculture (USDA), and they wanted information about breeding bees as a way to defend the United States from the invasion of the "killer" Africanized honey bees. I explained the mating behavior of queens and drones and gave details about instrumental insemination, how you ejaculate and collect the sperm from the males and inject it into the vaginal orifice of the queen. The day after it aired, I had phone calls from friends and family all over the United States asking me how I could stand there with a straight face and say all those words and give all those descriptive details. Sometimes reality is stranger, or more salacious, than fiction.

But the mating behavior of honey bees, as strange as it may seem to us, isn't fiction. The story began for our virgin queen several days before she took her first nuptial flight; it began with the song of the queen, singing out from the confinement of her wax prison, locked inside by her sisters. Her mother left with about 60% of the workers to establish a new nest, probably within a few hundred meters. Shortly after her mother left, perhaps several days ago, one or more of her virgin sisters emerged from their cells. They may have been mature, ready to emerge, for up to 5 or more days prior to finally cutting themselves free of their wax prisons, giving time for their cuticle—their exterior shell, the exoskeleton, their armor—to harden and for their stinger, a modified organ for laying eggs, to become an effective weapon. The worker sisters until now had been neutral, providing unlimited food to all of the perhaps as many as 25 developing queen larvae. The queen had determined their fate, worker or queen, long before she left, when she laid the egg in the queen cup, the nub of a nascent queen cell attached vertically to the bottom of a comb. However, now the workers assume a more prominent role in the succession of the nest and the formation of additional secondary swarms that might issue forth. Strong colonies that are congested with adults and full of brood frequently produce a second swarm and even a third or fourth, though the later, smaller afterswarms, usually tiny, have little hope of surviving.

Virgin queens are eager to emerge, but the workers hold them prisoner in their cells. The first one out has the upper hand in inheriting the nest, a precious inheritance of comb, brood, food, and workforce. She's more mature, harder, and stronger and will seek out the cells containing her rivals, chew holes in them, and sting her softer-bodied sisters sequestered inside. But the workers have other plans. The nest is very congested with bees, and the colony is primed to produce more than one swarm, so more than one virgin queen will be needed, alive. Workers control the emergence of queens by repairing the cells from the outside as the queens chew at them from the inside. Occasionally, they'll open a hole and feed the virgin inside, then reseal it. Virgins participate in the transition by pressing their bodies against their cells and vibrating their wing muscles, generating a sound called piping that has two forms, tooting and quacking. *Tooting* is a song of repeated syllables of 1 to 2 seconds that decrease in length with each one down to about 0.1 seconds; then they stop and start over. *Quacking* is like tooting but consists only of short 0.1-second syllables. The song is in the key of G sharp.

The workers allowed one virgin, her sister, to emerge before her but denied her access to the remaining queen cells, blocking her access. The free virgin ran from comb to comb, encountering cells with virgin queens that her instincts told her to destroy. Repeatedly being blocked, another behavioral program was inserted into her repertory, and she began tooting. The imprisoned virgins quacked back. The tooting of a virgin queen stimulates the colony to swarm yet again, the second swarm issues forth, again with about half of the remaining bees. Playing a recording of queen tooting to a colony that's queenless will stimulate the bees to leave as if they were engaged in reproductive swarming, another instinctive response. After the second swarm departed, the remaining workers released their prisoners, letting them cut their way from their cells. The first to emerge, our queen, then located the cells of her rivals, cut holes in them, and stung them to death through the hole in the cell wall using her curved, smooth stinger, unlike that of her worker sisters that's straight and barbed. As soon as she dispatched a queen, the workers moved in behind her and opened the cell, removed the body, and cast it out of the nest. During the confusion and frenzy of locating cells and killing, rival queens may have had the opportunity to cut free and begin their frantic search for others, each engaging the other in mortal combat until just one remained, reminiscent of the 1986 movie *The Highlander*, "there can be only one."

The songs of the queens are audible to the beekeeper walking through the apiary, but the workers only sense the vibrations in the combs. Bees don't hear sounds; they feel them, perhaps like Beethoven sitting deaf on the floor of his apartment working at his legless piano, feeling his music through the floor as he composed it. Bees sense sound through organs specialized for touch perception, the chordotonal organs located strategically at articulating junctions of their bodies. Many beekeepers have stopped, listened, and marveled at the orchestra of sounds emanating from their swarming hives, often baffled at their origin. This stage of seasonal colony development has been called *swarm fever*, a "disease" that spawns new colonies but represents a loss in potential honey and revenue to a beekeeper as the bees continue to fly away with swarms and afterswarms.

A new virgin isn't a worker and not yet a full queen. Her complement of pheromones, chemicals produced by glands in the queen that are released externally and affect the behavior of workers toward them, hasn't yet developed, informing the workers of her special status in the nest. This won't happen until after she returns from her mating flight and is inseminated and begins egg laying. And she hasn't yet acquired the full complement of "hive odor," an assortment of chemical cues that adhere to the surface of the body that will identify her as a hivemate. As a result, virgins are often treated aggressively by the workers, biting, tugging, and in some circumstances forming clusters around and immobilizing them. When trapped by overly assertive workers, the new virgin will begin piping her tooting song, which has the immediate effect of freezing the motion of the workers near her, allowing her to escape their grasp.

About 3 days after our queen emerged, she began showing the first signs of sexual maturity, opening the orifice to her sting chamber at her rear end. The workers began taking notice, touching her with their antennae and gathering around her, especially in the afternoon, the time for mating flights to take place. Ideal conditions for mating flights for queens is mid-afternoon, usually between 1:00 and 5:00 p.m., with the highest frequency between 2:00 and 4:00 p.m. Clear weather, temperatures above 20°C, and wind velocities below 19 kilometers (km) per hour are optimal. This also corresponds to the time that drones leave the nest and fly in search of queens. Drones become sexually mature 4 to 14 days after emergence, with a majority initiating flights outside the nest when they're 6 to 8 days old. Drones may take flight as early as 11:00 a.m. and not return until as late as 5:30 p.m., but the majority of them are active between 2:00 and 4:00 p.m., corresponding, as you might expect, to the peak times of queen flights. The only reason queens and drones fly is to find each other and mate, with the singular exception of swarming.

Now mature, the workers continued to harass her, especially in the afternoon, at times tugging and pulling on her, urging her toward the entrance. The queen may sing in response, then run around the nest making a whirring sound with her wings. Some bees begin making runs through the nest toward the entrance, appearing to encourage the queen to follow. Some bees join the chorus of the song of the queen, lining up at the entrance; exposing their orientation scent gland, the Nassanov gland, located on their back between the last two abdominal segments; and fanning their wings and creating a buzzing sound audible to the human observer, while providing a scent trail home for the queen, lest she gets lost. A major change occurred in the virgin queen's sensory system before she appeared on the entrance. Until now, she stayed in the dark, avoiding light, a response called *negative phototaxis*. Now, she's *positively phototactic*, attracted to light, drawing her to the entrance of the nest and outside. Now ready to mate, her path is blocked from returning to the nest; her only option is to take flight.

9.2 The Mating Flight

A virgin will normally take one or two short orientation flights, a common behavior of workers and drones that are taking to the air for the first time. Orientation flights are of short duration, lasting a few short minutes, during which a queen may turn and hover in front of the hive, learning landmarks and characteristics of the nest to aid her in finding the nest when she returns. After orienting, she'll fly off in search of places where drones hang out, called *drone-congregating areas* (DCAs). These areas have been known since at least the mid-19th century. A beekeeper in southern Brazil reported an observation in his apiary in 1849 where a virgin queen was pursued by a drone comet that varied between about 10 and 0.3 meters (m) from the ground. He estimated that two to 30 drones gave chase. In 1892, a beekeeper in the United States reported two places he had discovered where a large number of drones could be heard humming overhead. But it remains a mystery what defines their boundaries and characteristics. They exist year after year in the same geographical space, a space that doesn't seem distinct in any way; but the queens and drones find their way there, even though all of the memory of the sites died with the resident drones over the winter. DCAs tend to be located away from apiaries, away from hills and high trees, protected from the wind, and over open ground. However, there are many exceptions, including DCAs that have been found over forest treetops and water. A DCA is typically contained in a volume of space 30 to 200 m in diameter and 10 to 40 m above the ground, without physical boundaries but bounded nonetheless by features invisible to us. Drones tend to ignore queens, and few matings take place outside of this volumetric space. Drones may locate many DCAs, 10 or more, within a flight range of 5 km or more.

The quest for locating and mapping DCAs goes back to the early to mid-1960s by German and American bee biologists. DCAs were known from the very pronounced sounds that emanated from them, the droning wings of thousands of males flying above. Someone had the great idea that you could suspend a queen from a helium-filled balloon and attract males. Fly the balloons around an area, and find where the drones hang out. This method demonstrated two things: first, that the queens produced some kind of attractant pheromone and, second, that drones are only attracted to it within a confined volume of space, the DCA. Subsequently, after years of monitoring, it was found that the same volumes of space served as DCAs year after year.

As a queen enters a DCA, she's singing a new tune, this one with chemical communication, a major channel of communication in the insect world. Drones that are lured by the aphrodisiac begin to chase, forming a comet-shaped swarm, each scrambling to reach the queen. They join the chorus with the humming of their wings, audible to the human observer landlocked below, and by emitting their own chemical song that attracts others to join the orgy. As many as 20,000 drones, from more than 200 colonies, may attend the DCA.

It has long been known that queens mate while in flight, away from the hive. Anton Jansha, the imperial and royal teacher of bee culture to Maria Teresa, empress of Austria, observed in 1771 that drones and virgin queens took flight on fair and warm afternoons, after which queens returned with a mating sign, a visible structure at the tip of the abdomen. A few years later, François Huber, a Swiss naturalist, conducted experiments that clearly showed that queens and drones mated outside the hive and that the mating sign consisted of part of the genital organs of the drone. Prior to Jansha and Huber, there were a lot of strange ideas about how queens get inseminated and eggs get fertilized. It was generally believed that queens mated inside the hive with their own drones. Another view, expressed by Dutch naturalist Jan Swammerdam, was that the large number of drones within a colony emanated an odor that penetrated the body of the queen and made her fertile. When he dissected the males and found the huge internal genitals filling the abdomen, compared with the much smaller ovaries of the queen, he thought this was support for his hypothesis. Copulation looked to be impossible (Figure 9.1). Another common view was that the queen laid eggs, then the drones came along and deposited their sperm onto the eggs in the cells; again, why all those drones if they weren't doing something fertile in the nest? Then there was the view that queens didn't need to mate; they were self-fertile and didn't need males at all. This is a view that was partially correct and supported by the experimental results of Johann Dzierzon, who in 1845 demonstrated that males have no father but workers do, haplodiploidy, the basis of sex determination in the bees, ants, and wasps.

But this is just the beginning of the story of the discovery of honey bee mating behavior. The huge genitalia of the drones bothered morphologists. The huge endophallus (entomology-speak for penis), accessory glands, and testes fill their abdomen. How can the drones possibly use this to inseminate the queen? And how many times does a queen mate? It was generally believed that queens mated with just one male. Evidence for this came primarily from the occurrence of the mating sign when queens returned from their flights. It was believed that the mating sign blocked subsequent matings. Additional support came from microscopic studies where the number of spermatozoa were counted in the spermatheca of the queen, the organ where she stores the sperm, and counted in the amount of sperm a single male produced. They were equal, therefore, she mated one time. However, by 1932, there were doubts. W. J. Nolan, a USDA apicultural scientist, and one of the early developers of instrumental insemination technology, noted in 1932 the growing number of observations of double matings. These were observations of queens that returned from more than one mating flight with a mating sign. This motivated William Roberts, another USDA pioneer of honey bee genetics, to conduct intensive and continuous observation of 110

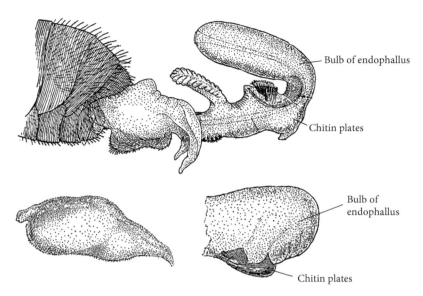

Bulb of endophallus

Chitin plates

Bulb of
endophallus

Chitin plates

Figure 9.1 Drone genitalia. Upper left: Fully everted penis. The entire mating apparatus (endophallus) is contained within the abdomen of the drone until ejaculation. The complex parts of the endophallus fit within the sting chamber and vagina of the queen, filling lateral pouches and penetrating to the opening of the median oviduct, thus firmly anchoring the drone inside of the queen until ejaculation. Upper right: Coagulated mass of mucus. This mass of mucus was removed from the vagina of a mated queen. It was deposited by a single drone during ejaculation. Bottom: Bulb of the endophallus taken from the sting chamber of a queen, the mating sign. This piece broke free of the endophallus of the drone and remained in the queen following mating. From *Anatomy of the Honey Bee* (p. 310, fig. 107), by R. E. Snodgrass, 1956, Comstock Publishing Associates. ©1956 with permission from Comstock Publishing Associates, a division of Cornell University Press.

queens, of which 55 returned twice with a mating sign, causing him to conclude in 1944 that two matings was not rare. He confirmed these results using color variants of workers and drones. He showed that the different colors of workers, yellow and black, found in his experimental hives must have come from at least two matings. But still, most believed that queens mated with one male on a given flight, with some exceptions. There were a few observations of queens returning from a mating flight with more than one mating sign. The sign of a previous male still attached to the sign of the last.

In 1958, Stephen Taber and James Wendel published a paper claiming much greater numbers of matings for queens, seven to 10. They based their conclusion on an analysis of hundreds of colonies from Canada and the United States where queens mated naturally with drones from colonies containing a genetic marker, cordovan, for color. The cordovan trait is recessive; a bee must be homozygous for the cordovan gene, have a copy from each parent, to express the color. The mutation changes the normal black color of the integument (called the *wild type*) to brown. Cordovan-colored queens that were themselves homozygous for the mutation took mating flights from apiaries that contained drones with both the cordovan and wild-type genes. Their worker progeny then would be of both types if the queen mated with males of each type: cordovan if their father was cordovan, not cordovan if their father was normal. The proportions of the cordovan and normal workers produced within the colonies were used in a mathematical model to estimate the number of different males each queen mated with, even though only two parental types could be distinguished. Such models are still used to estimate the average number of times queens mate, though there are now many more genetic markers used in the analyses; and the estimates exceed 20 males per queen.

What is remarkable is that the obvious study didn't take place until 1955! Professor Jerzy Woyke, a Polish apicultural scientist, dissected the oviducts from queens returning from mating flights. He then measured the volume of semen contained in them. (I consider Professor Woyke one of the six great honey bee scientists of the 20th century, along with Harry Laidlaw, Otto Mackensen, Roger Morse, Friedrich Ruttner, and Warwick Kerr.) It was known that a single drone produced an average of 2.2 cubic millimeters (mm^3) of semen; however, the oviducts of a returning queen contained up to 20 mm^3. Queens were mating with up to 10 males on a single flight!

Professor Norman Gary was my major advisor for my doctorate at the University of California at Davis. He was one of the pioneers in locating and mapping DCAs and discovered that a single compound, 9-oxo-2-decenoic acid, produced in the mandibular glands of queens, was the main attractant used by queens to lure in drones. The compound, known as "queen substance" or "queen mandibular pheromone," had been previously discovered and its effects on worker behavior were already well known but not its role as a "sex attractant" pheromone. Initially, he flew queens tethered to a line suspended in the air by a balloon. He carefully mapped a large area around the university at Davis and found many DCAs, one about 500 m from the bee biology

research laboratory. There, he built two towers, between which he suspended a
cable. The cable could be raised and lowered; drones and queens fly at different
altitudes depending on wind velocity. On this cable he suspended virgin queens
and observed and filmed the behavior of the queens and the drones that were
attracted to her. He had the queen substance synthesized by a professional lab,
then suspended lures containing only the pheromone and studied the effects of
the pheromone without the visual stimulus of the queen, confirming that the
single compound queen substance was sufficient to attract drones. With many
hours of observation, he provided this description (Figure 9.2):

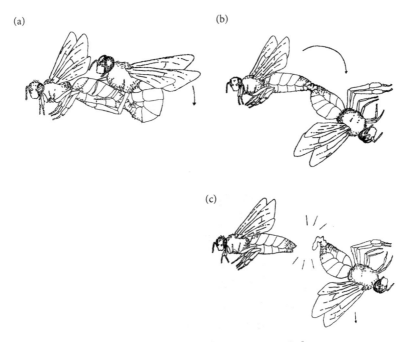

Figure 9.2 Illustration of a queen honey bee mating. a: A drone mounts
the queen while in flight in a drone-congregating area. b: After everting his
endophallus into the queen, the male deposits his semen followed by a mass of
mucus, then falls back paralyzed. c: The bulb of the endophallus breaks off and
remains in the queen, forming the mating sign; and the drone falls to the ground
and dies. Paralysis occurs when the abdominal muscles contract, squeezing
the abdomen dorsoventrally (compressing back to belly), creating hydrostatic
pressure that forces the endophallus out of the abdomen and into the queen and
powers an explosive ejaculation. From "Activities and Behavior of Honey Bees,"
by N. E. Gary, in J. Graham (Ed.), *The Hive and the Honey Bee* (p. 353, fig. 38),
1992, Dadant and Sons. Reprinted with permission from Dadant and Sons.

The drone alights on the back of the abdomen of the queen, grasping her with all six legs, his head extending over the thorax. The abdomen of the drone curls downward until it contacts the tip of the queen abdomen. If the queen opens her sting chamber, the penis everts and ejaculation occurs very rapidly. If the queen fails to open, the drone may remain in this position for several seconds or until another drone knocks him off. As soon as ejaculation occurs, the instantly paralyzed drone releases from the queen and topples over backwards, and two to three seconds later a distinct "pop" is heard as the two separate. If this explosion fails to occur, they may fall to the ground where they separate within a few minutes. Drones are strong fliers and are capable of carrying the queen along with them in flight. (Gary, 1992, p. 352)

The actual mechanics of mating, the positioning of the drone and queen and the manner in which the endophallus, internal in the drone, everts and fits into the anatomical structures of the queen, were still unclear. Hypotheses were that the mating position was face to face, with the drone on his back beneath the queen, or that the drone mounted the queen from behind, on top of her. The "male below" hypothesis was favored because of the size and shape of the penis of drones that had ejaculated and were found lying on the ground after mating. Both hypotheses assumed the drone was the stronger flyer and carried the couple through the air. The descriptions and films of Gary taken of his queens tethered between his towers at a DCA in the early 1960s resolved the positioning question. The queens that he studied were immobile; they weren't flying. And they weren't cooperative. They didn't open their sting chambers revealing the vagina and vaginal orifice, giving the drones access. To get mating behavior, the chamber had to be propped open with a metal O-ring, removing any natural behavior or control the queen might have. Drones could mount and evert their genitalia into the queen, but most became stuck and couldn't disengage; therefore, observations of the entire process of subsequent males removing the mating sign and the termination of copulation weren't "natural," although, Gary did observe one of his tethered queens mate with 11 drones in rapid succession.

The queens on tether were suspended stationary on a cable between towers. Professor Gudrun Koeniger wanted to observe more natural behavior from the drones as they pursued, overcame, and mated with a flying queen. This was the early 1970s, and there was still no way to follow and observe an actual flying queen. Koeniger built a tower at a DCA and at the top mounted a motor and attached a fixed rod that would rotate 360 degrees parallel to the ground. At the end of the rod she fixed a virgin queen. On the rod was mounted a motion film camera focused on the queen. As the rod circled at the top of the tower, the queen moved through 360 degrees with the camera filming her. When flown during the afternoons when drones normally fly, comets of flying drones would form and chase the queen in a circle while Koeniger filmed them. She was able to observe how the drones chase and catch the queens, mount them, and mate. Most of the matings were incomplete due to the necessity of artificially propping open the sting chamber and the lack of cooperation of the queen. However, she was able to study the incomplete eversions of the failed matings and learn about the mechanisms of mating in the queens and drones.

Most believed, like Gary in the quotation, that the strong robust males must carry the queen through the air while they mate. But that's not likely when you consider that the drone is paralyzed. In 2012, a Swiss nature film company produced a documentary called *More than Honey*, a film about the beekeeping industry in California and the decline of honey bees. They decided to tackle the centuries-old mystery of the mating of the queen. Still no one had observed it in a natural state, and so many questions remained; or at least the observations of Gary and Koeniger needed to be substantiated. They equipped remote-controlled mini-helicopters with high-speed cameras, chased virgin queens on mating flights, and joined the drone comets in the DCAs, drones chasing drones. They captured for the first time the natural mating of drones and queens while in flight. The film is nothing short of amazing, it can be found on YouTube and all over the Internet; just search for "honey bee mating flight" or "more than honey." And, by the way, the queen carries the drone through the air while he's flopped backward paralyzed. Now the story is complete (Figure 9.2).

As a queen flies through the DCA, each mate deposits his sperm into her vagina and median oviduct. The median oviduct is a thin tube that opens at the front end of the sting chamber, the cavity just forward of the rear tip of the abdomen where the stinger is retained when retracted. It appears that the queen must cooperate. She must open her sting chamber for him to gain admission and lower a structure called the *valvefold* that lies between the vagina and the median oviduct (Figure 9.3). The valvefold was discovered by Professor Harry Laidlaw in 1939 as part of his doctoral research at the University of Wisconsin. It was the biggest anatomical impediment to

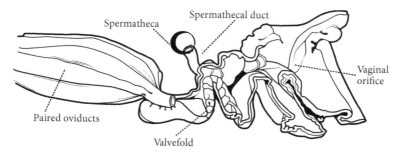

Figure 9.3 Reproductive tract of a queen honey bee. The posterior (rear) of the queen is right; anterior (front) is left. The drone inserts his genitalia into the sting chamber, then explosively ejaculates into the median oviduct. The queen lowers the valvefold to allow the semen access to the paired oviducts. The paired oviducts fill with sperm from the many mates of the queen, each subsequent male pushing the semen of the previous males forward. After mating, the queen returns to the nest where, over a period of hours, she presses the sperm in the oviducts back toward the sting chamber, perhaps like squeezing a tube of toothpaste. As the sperm are pressed past the spermathecal duct, they may swim up the duct and into the spermatheca where they're stored, sometimes for several years. From *Queen Rearing and Bee Breeding* (fig. 77), by H. H. Laidlaw and R. E. Page, 1997, Wicwas Press. Reprinted with permission.

developing the technology and techniques for instrumental insemination of queens. The male's genitalia extend into the sting chamber and forward into the vagina, followed by a forceful ejaculation caused by the rapid constriction of abdominal muscles producing a high hydrostatic pressure, first releasing semen, that's propelled into the vagina and median oviduct, followed by a large quantity of mucus deposited behind it in the sting chamber. Plates consisting of chitin, material that makes up the exoskeleton, are located on the bulb of the endophallus of the genitalia. They peel off and, along with the mucus deposited, form the mating sign that plugs the opening of the vagina and sting chamber. The drone falls backward, dislodging his genitalia, and falls to the ground. The queen is left with a load of semen in the median oviduct and the mating sign stuck in her sting chamber (Figure 9.3).

My former graduate student David Tarpy studied the mating behavior of 32 queens. Of the group, 30 mated successfully; and of those 30, 22 began egg laying soon after flying. The remaining eight left on the next day to take another mating flight, but David captured them when they tried to leave and didn't allow it. They were confined to the hive and after a short time began egg laying. He sampled worker progeny of each of the queens and, using genetic markers, was able to determine how many different fathers they had, and therefore the number of males with which the queens had mated. He found that his queens mated with up to 13 males on their first flight, an average of five. The queens that attempted to mate again also had an average of five. The number of times they mated had no effect on their decision to fly again. They weren't counting the times they mated. He timed the mating flights of queens and found no difference in the time spent between those that ceased mating flights and those that attempted a second. There was no relationship between flight time and the number of matings. Two queens failed to return from their mating flights, constituting a failure rate of about 6%.

Professor Woyke showed that queens that receive 5.3 million sperm or more on their first mating flight aren't likely to take another. After dissecting and carefully studying many mating signs, he also concluded that the mating sign is usually composed of the genitalia of two males rather than one. He suggests that queens terminate a mating flight when they get two mating signs stuck in their sting chamber, which prohibits further mating, a matter of chance.

Males fall to the ground paralyzed and die after depositing their semen and the mating sign. Workers at the entrance carefully inspect a returning queen and take especial notice of the mating sign, while the queen walks around dragging the tip of her abdomen on the nest substrate, dislodging it. Workers will pull on it and remove it if she's unsuccessful. Workers continue to act aggressively toward her until she begins egg laying, so the following day they may again push her toward the entrance and send her on another mating flight. A queen may take up to three mating flights where she may mate with none or up to 13, or more, males on each flight. Once she begins egg laying, she has her full complement of pheromones, the scent of the queen is complete, and the workers accept her as a full queen.

9.3 Acquiring and Using Sperm

The question that remained was how the queen used all of the sperm she collected from the DCA. The sperm of just one male could effectively fill her spermatheca. What happened to all of it? This was actually two questions: How does she fill the spermatheca? And how does she ration it out to eggs? The use of the stored sperm provides some insight into how the spermatheca is filled.

The spermatheca of a queen is a very small sphere full of fluid. The spermatheca of a virgin queen is clear, translucent. That of a mated queen is more like an off-colored pearl, opalescent with waves of light brown marbling, the color of the semen produced by drones. It looks like some tiny planet from a different solar system. It's easy to determine if a queen has mated by extracting her spermatheca and looking at it; it can be done with the unaided eye. But I don't recommend this as a way of validating if your queen has or has not mated; of course, the queen would be dead. Because of the "whorls" of sperm visible in the spermatheca, many believed that the sperm from the different males must enter the spermatheca together and stay together in clumps inside. Stephen Taber showed that the sperm of many different males gain entry into the spermatheca and that at least two are used by most queens at any given time. He also looked at the patterns of sperm use over extended periods of time by observing how the proportions of cordovan and wild-type (normal) offspring changed in cohorts of emerging new bees. For instance, one set of samples of bees emerging over 24 hours might be 60% cordovan and 40% wild-type. At another time they may be 60% wild-type and 40% cordovan. Clearly, the sperm used by the queen to fertilize the eggs wasn't random. He concluded that the sperm didn't mix "appreciably" before or after entering the spermatheca. However, as an ever-skeptical graduate student, I looked at his data, which were substantial, and believed that his data didn't support his conclusion. Yes, the sperm weren't randomly mixed, but they did a fairly good job of mixing. I thought about how hard it would be to mix, say, 10 different-colored packages of spaghetti in a pot. I doubt that any amount of mixing you and I were willing to do would ever create a random mixture.

I wanted to do a better, more complete study of sperm mixing, so I enlisted three of my best friends, Harry Laidlaw, Bob Metcalf, and Bob Kimsey. Bob Metcalf was a new breed of molecular biologist, employing genetic biochemical markers to look at genetic relatedness and behavior of animal populations. He had a lab set up for the molecular studies I wanted to do and trained me to do them, with him. We looked at colonies where the queens had mated with multiple males with different biochemical markers and followed them over time. We could distinguish three different mating types, one more than Taber. Our results were similar to his, but we concluded that the sperm were fairly well mixed, considering the challenge.

Bob Kimsey was a graduate student in entomology, one of my cohort. He knew how to do histology, prepare tissues like a spermatheca, slice and section them, and do photomicroscopy, make photographs through a microscope. Professor Laidlaw instrumentally inseminated a set of queens for us. Then, at time intervals of 15 minutes, 1 hour, 2 hours, 4 hours, and 12 hours, we dissected out the spermathecae of queens; and Kimsey fixed and sliced them into thin sections for microscopic examination. We were observing the process of the spermatheca filling with sperm. This is a process that takes up to 24 hours. We showed that 15 minutes after insemination the spermatheca is mostly empty, but over time it gradually fills with sperm as they migrate up the spermathecal duct. As the sperm enter, they disperse; they don't enter in clumps or aggregations but simply diffuse throughout whatever space is available. It's likely that the sperm that enter last have less space in which to disperse and may be over- or underrepresented in areas of the spermatheca, perhaps contributing to the observed temporal fluctuations in workers with different fathers.

The final experiment was done with Professor Laidlaw. It required someone with lots of experience at instrumental insemination, and since he was one of the early developers of the technology, he was the obvious collaborator. Harry inseminated queens with sperm from males with six distinguishable genetic traits. These traits were derived from two different mutant genes that affected the color of the bee (cordovan or wild type) and eye color. The eye color gene could produce three distinct colors of eye, tan, red, or wild type, depending on the specific form of the gene used for the insemination. The worker progeny of the queens was of six distinct types. The order of insemination of the semen of the males having the different traits was controlled by collecting the sperm from the different males into the syringe in serial order to look at the degree of sperm mixing that might take place as the spermatheca filled. Some inseminations took place on subsequent days, again with different genetic markers, to look at the effects of late matings on the proportion of sperm that enter the spermatheca. Workers that emerged from the combs of colonies with the inseminated queens were collected over a period up to 2 years. The results were clear. Sperm do mix appreciably as they fill the spermatheca, though not completely randomly. The order that the sperm go into the oviducts doesn't affect how many of them get into the spermatheca. The males that contributed last had as many as those that contributed first.

Therefore, the filling of the spermatheca didn't affect the ability of the later-arriving sperm to get in and get distributed. Even sperm injected several days after the queen had been inseminated repeatedly with more than enough sperm to fill the spermatheca were represented in the spermatheca, though at a lower frequency (Figure 9.4).

How does the queen mix the sperm? Obviously, she does. As the sequential drones mate with the queen, they deposit their sperm into the vagina

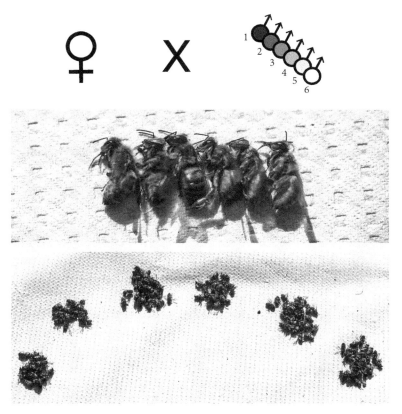

Figure 9.4 A single queen was "mated" by instrument to six different drones (upper diagram). Each drone carried a unique combination of visible mutations for eye and integument color, giving rise to six distinct worker phenotypes corresponding to each drone father (middle panel). The bottom panel shows the worker representation of all six drone fathers after emerging from a single brood comb for several hours. Reprinted with permission from *The Spirit of the Hive: The Mechanisms of Social Evolution*, by R. E. Page, Jr., Harvard University Press. Copyright © 2013 by the President and Fellows of Harvard College.

and median oviduct. With his explosive ejaculation, each subsequent drone pushes the sperm loads of the previous males further and further into the lateral oviducts, filling them. When the queen returns to the nest, she removes her mating sign, then spends the next 12 to 24 hours processing the sperm she's collected. She ensures a mixture of sperm by passing the sperm back toward her sting chamber like squeezing a tube of toothpaste. She can be seen pulsating and contracting her abdomen. Her oviducts are covered with a thin layer of muscles that can contract and push the sperm back. The sperm pass by the spermathecal duct. They only have access to the spermatheca when they're close by. They swim up the duct; she doesn't press them in. The sperm from each male probably passes the duct in serial order, still being clumped together; but as they swim into the spermatheca, they mix. Once the load of one male is passed into the sting chamber, it dries into a thread and the workers pull it out of the tip of the queen's abdomen and remove it. This is a very effective way to ensure mixing. All the queen has to do is have some steady rate of pushing the sperm back, and all of her mates will be represented in the spermatheca and in her progeny. But this is very inefficient! She takes in 10 to 20 times more sperm than she needs (Figure 9.3).

Why is the transfer of sperm from mating so inefficient? One view is that it's a consequence of the egg-laying equipment of the queen. It makes egg laying efficient at the cost of sperm use efficiency. However, other social bees such as the stingless bees only require the sperm of a single male to fill the spermatheca with as many sperm as a queen honey bee gets. This is true also for some species of ants that build big colonies and the queen lays lots of eggs. Another view is that the queen needs to mate with a lot of males because one male can't produce enough sperm to fill her. But the examples given here block that view as well. And if you look at the size and complexity of the drone's genitalia, it doesn't look like there's a limit to what he could produce! Most likely the inefficient mechanism of sperm acquisition is a result of the need to produce workers with lots of genetic variability. Increasing numbers of matings increase variability. Look at what queens go through to get genetic variability in their offspring: (1) they avoid inbreeding (it reduces genetic variation) by not mating with their brothers, who are plentiful in the nest; (2) they fly up to several kilometers to encounter large numbers of unrelated drones; (3) they have evolved a unique reproductive anatomy to ensure the capture, mixing, and sequestration of sperm from many, many males; (4) they nurture and keep the sperm alive for up to several years to avoid the necessity of taking additional, potentially dangerous, mating flights in the future; and (5), something I haven't discussed in this book and won't, honey bees have an incredibly high rate of genetic recombination when they produce eggs. Recombination takes place during meiosis when the egg genomes are formed. Higher recombination produces more genetic variation in the eggs.

Colonies with more genetic variability within the workers survive better and produce more successful new queens and drones than those with less. When you study the individual behavior of workers within a colony, you find that not all bees do the same thing, not all progress through their short lives of 3 to 6 weeks following the same sequence of jobs performed in the nest, and not all of them make the same kinds of foraging decisions: when to start, what to collect, where to go, whether to recruit more foragers or not. It was long believed that this kind of interindividuality of behavior was derived from each bee having a unique set of experiences as she ages. Each is born at a different time, emerges as an adult in a different part of the nest, then encounters different combinations of stimuli as it matures. It was believed that workers are born with a clean behavioral slate, each capable and equally likely at birth of becoming and doing anything. This is a view that we generally take with our own offspring, or at least we like to believe that at birth they could each become a bank robber or US president. My two children, though they have the same mother and father and, as a consequence, share half of their genes in common, could have been born on different planets with different sets of alien parents, if you look at their behavior (fortunately, no bank robbers). However, over the last 30 years it has become accepted that much of the variation we observe in behavior within colonies is due to genetic variation, and some of that's due to different workers having different fathers who contributed different genomes. The actual total of direct genetic effects on foraging behavior is less than 10%, so environment and experience are still the most important determinants.

When you look at colony responses to certain important environmental challenges, those with a greater mix of fathers contributing to the worker population survive better. This is true for resistance to disease, thermoregulation, and establishing a new colony that builds up over the summer and survives winter. Mating with a large number of males reduces the variation between colonies, making each one more similar. By mixing the genes of the population in the spermatheca, each colony becomes more like a historical, statistical average of the evolutionary trajectory, effects of the environment, on the population, better adapted to cope with environmental challenges similar to those that occurred in the past. The colony resists system failures and survives better.

Drones are just vehicles to deliver their sperm package to the queen. They're weaponized delivery systems, armed at many levels for intense competition. They're large, robust, and strong flyers; natural selection has been extreme, adapting them to be competitive against the huge number of males produced by populations each generation (20,000 per colony). The chance that any male will mate successfully is perhaps two or three in a thousand, counting swarm and supersedure queens. Drones are armed with a sophisticated sensory system tuned for locating DCAs and finding queens. They accomplish this with large eyes, each with about 8,600 facets (each like a small eye) tuned to detect motion, like a flying queen, compared with 4,500 for queens and 6,900 for workers. In addition, their antennae are loaded with about 18,000 chemoreceptors tuned to the mating pheromone of the queen. Queens have about 1,600 and workers 2,600 receptors on each antenna. Their robust, strong bodies and long legs help them outcompete other males flying in the DCA as they grab and mount the queen and knock off other males. Their reproductive organs and genitalia are enormous, shaped by millions of years of competition for mates and competition for as much of their sperm as possible to be included within the spermatheca of the queen and to exclude as much as possible the sperm of other males. Their ejaculation is explosive, displacing the sperm of previous males further back in the oviducts, further from the sweet spot for sperm migration into the spermatheca that lies just below the spermathecal duct. The large genitalia are tailored to fit tightly inside the queen, with special structures to anchor it and to deliver the semen into the vagina and median oviduct, to deliver massive amounts of rapidly coagulating mucus into the sting chamber to plug the entrance and deny access to other males. The mating sign left behind is a remarkable structure formed from a huge amount of mucus and parts of their genitals that broke off and were left behind (Figure 9.1). Part of the genital apparatus of the drone is a hair brush that fits under the mating sign of the previous male and removes it as the drone everts, clearing the way for his own delivery.

One can imagine an evolutionary arms race taking place over time where drones are selected for two things: blocking access to mating with the queen and removing the mating sign so that they can gain access. Some argue that the mating sign is a structure designed simply to block the escape of semen from the oviducts until the queen returns to the nest and begins processing it—like plastering up a hole in the wall to keep out the cold and damp.

However, the plug does in fact still function to block additional mating, as pointed out recently by Professor Jerzy Woyke. It's likely that most queens return to the nest from mating flights because a male deposited a plug that couldn't be removed by the next male. The presence of double mating signs and sperm deposited on the backside of a plug aren't uncommon, showing a degree of effectiveness in excluding other males. The arms race continues. I wonder what new features the complex genitals will have a few million years from now.

The queen isn't a passive player, manipulated by the urges of the males and left out of the evolutionary arms race brokered by natural selection. The queen is designed for more than the mating function; she must also lay eggs, coordinate some of the social life of the colony, and swarm the next spring. Her reproductive tract reflects those needs. However, the shape and organization of her sting chamber reflect her reproductive interests of accommodating the evolving, huge reproductive structure offered by the males and guaranteeing that individual males can't dominate her reproductive interest of high genetic diversity of her offspring. Her sting chamber contains pouches and stiff hairs that provide anchor points for the male to form a firm grip during sperm transfer. Her oviducts, location of the spermathecal duct, and vaginal orifice are part of her sperm-mixing machinery. Her curved sting is designed to be pushed up and out of the way within the sting chamber by the genitalia of the drone and the mating plug. Duels between the stinger and genitals can occur, resulting in the endophallus of the male being skewered and fatally damaged and failed mating success for both. She's a strong flyer, able to carry the weight of the drone, paralyzed in the climax of his life. And her valvefold is the access door for ejaculations; she has control. The drones and queens evolved together to find a truce solution; though their interests are very different, it's one that works, for now.

Gregor Mendel is known as the father of genetics, though the term was actually coined by William Bateson in 1905. In 1865, Mendel published his famous paper that established the two fundamental laws of gene transmission: the independent assortment and segregation of traits. Mendel was an Augustinian friar and abbot in what's now the Czech Republic. He spent 7 years studying the common garden pea, discovered his laws, and published his seminal work in 1866. He spent the rest of his life trying unsuccessfully to breed a better bee. He failed because he couldn't control their mating, something he recognized as his main obstacle. He built large enclosed structures for queens and drones to fly, hoping he could get them to mate but to no avail.

He wasn't alone. Attempts to control fertilization of eggs go back to at least the early 1700s, but it wasn't until the 20th century that the cumulative knowledge of basic anatomy, morphology, and behavior was sufficient to develop instrumental technology. It was primarily the efforts of Lloyd Watson, W. J. Nolan, Harry H. Laidlaw, and Otto Mackensen that led to the instruments and technology we use today. Their contributions opened up the honey bee to the breeding techniques that had been so effective at improving plant production and scientific studies of genetics, behavior, anatomy, and physiology. Today, we have many new genetic tools to work with including a complete sequence of the honey bee genome, ways to reduce the expression and effects of specific genes, and, most recently, the ability to edit the genome of the honey bee, making changes in the 10,000 genes of the 236 million base pairs that reside on the 16 chromosomes. Because of the efforts of these researchers, and others, the future of honey bee science is bright.

9.4 The Song Ends

The queen flies through the air conducting an orchestra of drones with her scent, her conductor's baton; the drones respond with their wings, the buzz of the song, punctuated with the popping crescendo of successful ejaculations, followed perhaps by the gentle beats of their paralyzed bodies contacting the ground. The queen returns to the hive after the last mating flight, the royal succession complete—the song has ended. Her worker sisters welcome her return; the transition from one generation to the next, the old superorganism to the new, has begun. Her ovaries and spermatheca contain the sequestered germ line from which the new organism will develop. She re-enters the dark confines of the hive, photonegative again, hiding from the light, and becomes the servant of the workers until the next year when she sings again as she prepares to swarm and orchestrates the orderly transition of power and the transition of the swarm that accompanies her to her new home.

9.5 New Life Begins

Deep within the darkness of the nest, the queen inserts her head into the carefully prepared wax cell on one of eight combs, one of 100,000 cells in a natural nest. Nurse bees have finished polishing the bottom and sides with their proboscis, part tongue and part straw, making this cell ready to receive a single egg. She inserts her forelegs into the cell, measuring its width. The width will determine whether she lays a fertilized egg or denies the egg access to one of the 6 million sperm lying dormant in her spermatheca, like futuristic space travelers. Next, she bends her abdomen under her body, toward the cell, inserts it, and deposits a single egg at the bottom—an individual act, a singular moment, an apparently personalized birth. But this act will be performed more than a thousand times each day, each time exactly the same. What appears to be personal is really the workings of an automaton, mindlessly repeating the same movements with the same result endlessly, day after day. Between egg-laying bouts, the queen is fed mouth-to-mouth by her worker entourage—protein to make more eggs. She's an egg-laying machine, a slave of the workers, domesticated by her own daughters.

If the cell is the width of a worker, the queen fertilizes it by releasing a small drop of fluid from the spermatheca, each drop containing a few to many spermatozoa aroused from their induced coma. One end of the egg has not fully formed, leaving an open netlike matrix, the micropyle. This end of the egg is pressed up against the spermathecal duct, a tube running from the spermatheca to the oviduct down which the sperm-containing droplet will pass. The droplet is deposited onto the micropyle, and the sperm cells begin the wiggle race to be first to penetrate the egg and fuse with the nucleus. A zygote is formed. The other sperm die and disintegrate, while the newly formed zygote divides. The micropyle closes, sealing in the contents of the egg. New life begins.

Drone-sized cells are larger than worker cells and receive an egg that's not fertilized. The queen can control the fertilization of each egg she lays and therefore the gender of the individual that will develop. When a queen is put on a comb that has been altered into a checkerboard of drone and worker cells, she works across the comb, laying only worker-destined, fertilized eggs in the worker cells and unfertilized, male-destined eggs in the drone cells. Professor Niko Koeniger, in a clever experiment, demonstrated how the queen measures the cell size with her front legs. He attached small tape "flags" on the outside edges of the front legs of a queen. He then placed the queen on drone-sized cells where she attempted to measure the cells, perceived them to be smaller than they actually were, and laid fertile eggs as if she was on a worker comb. The bees that emerged from the cells were indeed workers. Without the prosthesis, only drones were produced.

The egg begins its development, an embryo contained in a soft shell. The embryo develops inside the chorion (shell) of the egg for 3 days before the larva hatches. *Hatching* is a very loose term for what actually occurs. The chorion dissolves, and the larva eats it as it's "born." Nurse bees begin offering small droplets of royal jelly to the larva even before it's fully out of its protective envelope. Once released, the larva begins its five stages of development. Each stage takes roughly a day, terminating with the larva shedding its outer skin, integument, and growing a new, larger one to accommodate its rapid growth. Even though the larvae look soft to us, they're restricted to living inside a shell (cuticle) that must be replaced with each molt. During the fifth larval stage, the larvae are fed royal jelly by the nurse bees. The sugar content of the food and the amount fed change as the larva ages, regulating the development. The total time for larval development is about 6 days, ending with a final molt to the pupal stage. Prior to pupating, the adult workers cap the cell with porous wax—pupae need to breathe. Pupae don't eat; they just lie in their capped cells for 12 days, inside a cocoon they spun as larvae just prior to pupating, and go through a complete transformation from a soft, featureless blob into a wonderfully specialized adult worker honey bee. Twenty-one days after the egg was laid the new adult worker chews her way out of her wax nursery and enters a new world, ready to fulfill her part of a social contract that binds the colony together. She becomes a small part, the first cell, of something much larger and more complex than she, something that's organized for reproduction, nutrition, and protection with coordinated action of its many parts, capable of responding to its environment, containing a collective memory, and capable of making decisions—a superorganism.

9.6 Parting View

The extract of the poem at the beginning of this chapter was from E. B. White and was published December 15, 1945, in the *New Yorker* in response to a recent news release announcing the new and effective instrument developed by Professor Harry Laidlaw for inseminating honey bee queens. The poem that follows was Harry's response.

Response
The queen bee is a giddy thing
She plays, she romps, she takes the wing
She darts among the sun-lit hives
Without a thought of being wise.

Her offspring slave throughout the day
They feed her children as best they may.
They would like to see a movement
Directed toward stock improvement.

Harry H. Laidlaw, *San Francisco Chronicle,*

December 5, 1948

Epilogue

Readers of earlier drafts of this book suggested that it needed an epilogue to link the views presented here with human societies, to explore what we can learn about ourselves from the bees. I've resisted making such connections, leaps of faith, although metaphorical links between human and insect societies are obvious. While writing this book I struggled to find what I might consider to be universal or unifying concepts to explain the evolution of similar social structures of humans and insects. To explore this, you need to look for global properties of organization rather than specific evolutionary links, move from a view of meters to a view from the moon.

Members of complex societies live close together in closed nests, shared home sites, villages, etc., or in closely connected nomadic tribes. As groups, they typically have a set of tacit rules by which they live that involves working for the good of the group, systems of group and resource defense, internal mechanisms of policing cheaters that don't cooperate and live by the rules, a division of labor often associated with group defense and gathering and sharing resources, and usually asymmetries and rules associated with reproduction. These same general characteristics seem to apply broadly across eusocial insects (aphids, termites, bees, ants, and wasps), eusocial rodents (naked mole rats), higher apes, and humans. Why? The similarities are inescapable due to the nature of social contracts; they must have specific elements to protect the power and will of individuals, whether citizens of the United States of America or workers in a honey bee colony. The contract binds individuals to a society, but the specific social organization evolves by reverse engineering. Natural selection acts on the whole colony; social structure evolves to fit the needs of the group within a given environment.

Common evolutionary links between human and insect social behavior not only are unlikely due to more than 400 million years of evolutionary separation but also would be impossible to trace. The evolution of social behavior hasn't been progressive, ever-advancing along linear trajectories of more and more social complexity. As Darwin pointed out, evolutionary trees are bushes; some branches representing lineages are long and still growing, while others are short, with many dead ends. Evidence from bees shows the same patterns for the evolution of sociality. It's a history of sociality advancing along a particular lineage, then stopping, or even retreating, becoming less social, erasing evidence, and returning to solitary.

When we look at the genes themselves, we get a different picture. We have a picture of the same genes being around since quite soon after the beginnings of life, around 3.8 billion years ago. We share many genes in common with bacteria, roundworms, and insects. In some cases whole gene families, like the genes involved in insulin signaling, are structurally similar and have similar functions in nematodes, bees, and us. This paints a view of the evolution of life on earth as one where early in its development the gene sets we see today were tested, found to work for specific functions, and used. Genes encode specific proteins. Proteins become structural parts of bodies, drive metabolic systems, transport things into and out of cells, and facilitate many kinds of reactions and interactions of other proteins. In order to function, they need certain physical properties. Perhaps many of the countless different possible proteins that could exist were explored early, and we're left with a subset that meets the unique requirements to be useful for life on earth. But it's a long way from sharing genes in common with insects, a result of a relatively small subset of all the possible proteins they code and a singular origin of life, and sharing a common evolutionary history for complex social behavior.

Honey bees look like little people with frozen faces, staring at us from the entrance and top bars of a hive. It's easy to believe that they, like us, plan their future, feel satisfaction in caring for the family, love their queen, hate their enemies, and have emotional highs and lows with good days and bad. To view them in that way has a term, *anthropomorphic*, or *anthropocentric*, meaning like us, or human-centered. There are many pitfalls associated with this kind of thinking in behavior, even though it's often done. I am guilty in this book, sometimes lapsing to explanations of behavior using terms that most likely represent traits that are unique to us, shorthand to make it understandable to non-specialists. Anthropocentric thinking can obscure the way we view nature and lead to false conclusions. Look at Aristotle and honey bee division of labor: For more than 2,000 years it was thought that the bees that work in the nest were postpubescent old men because they're hairy! In fact, the older bees forage and aren't hairy because the hairs break off as they age.

Many years ago, in the 1970s and 1980s, when the field of behavioral ecology was young, behavioral research shifted from trying to understand the mechanisms of behavior to studying its adaptive (evolutionary) significance. The questions shifted from how animals do certain things to why they do them, a shift from trying to understand the operations of sensory systems, central nervous system integration of information, learning, and neuromotor outputs that result in behavior to understanding the ecological conditions that shape behavior. I remember some of the names given by behavioral ecologists to the observed behavior of birds and mammals, like *adultery, divorce*, and *sneaker*, names to describe alternative strategies to get mates. But the one that I found most incredulous was the use of the term *rape* for insect mating strategies. I remember the outrage from various sectors of academia: behavioral biology, anthropology, psychology, and even philosophy. How could this heinous behavior be attributed to an insect with all of the personal, deeply emotional, devastating effects that are linked to it?

I now see my work in a new light; we aren't so different, bees and humans. The elements of our social structures, and how they come about, have many similarities. Our individual behavior has been shaped by selection on our social structures over thousands of millennia, shaping us to fit the structures that are optimal for the environments we're in. Our adaptive responses to changes in our environment are likewise similar. When resources are abundant, bees are very docile, colony defenses come down, and nestmate recognition at the entrance gets loose, with foreign bees gaining easy access. Foragers come back from foraging trips full of pollen and nectar, lumbering in the air near the entrance in what looks like slow motion, reminiscent of an Airbus 380, the huge double-decker passenger plane, hanging in the air as it approaches the runway. You can open the hive without your veil, leave the hive open without fear of initiating robbing behavior from other colonies, and spend a laconic afternoon communing with your bees. But when the nectar in the flowers dries up, colonies transform into militarized societies. The defenses go up, the "fear" of external threats. Immigration services tighten up the entrance; their border control, guard bees increase in number; rejection of bees becomes more common; nestmate-recognition credentials are demanded, sometimes even rejecting their own nestmates—some might smell or appear like they don't belong—in their overzealous desire to keep out unwanted immigrants. And scouts go out on reconnaissance ventures seeking weaker colonies that they can exploit and rob. Maybe there are some lessons here.

But only humans have sociobiologists and political philosophers to ponder the biological and rational bases of our societies.

Resources

General Sources

Graham, J. (Ed.). (1992). *The hive and the honey bee.* Hamilton, Illinois: Dadant and Sons.

Page, R. E., Jr. (2013). *The spirit of the hive: The mechanisms of social evolution.* Cambridge, Massachusetts: Harvard University Press.

Seeley, T. D. (1995). *The wisdom of the hive: The social physiology of honey bee colonies.* Cambridge, Massachusetts: Harvard University Press.

Snodgrass, R. E. (1956). *Anatomy of the honey bee.* Ithaca, New York: Comstock Publishing Associates.

Von Humboldt, A. (2014). *Views of nature.* Chicago, Illinois: University of Chicago Press.

Winston, M. L. (1987). *The biology of the honey bee.* Cambridge, Massachusetts: Harvard University Press.

Chapters 1 and 2

Coevolution of Flowering Plants and Bees

Canto-Aguilar, M. A., & Parra-Tabla, V. (2000). Importance of conserving alternative pollinators: Assessing the pollination efficiency of the squash bee, *Peponapis limitaris* in *Cucurbita moschata* (Cucurbitacea). *Journal of Insect Conservation, 4,* 201–208. doi: 10.1023/A:1009685422587.

Cardinal, S., & Danforth, B. N. (2013). Bees diversified in the age of eudicots. *Proceedings of the Royal Society B Biological Sciences, 280,* 20122686. doi: 10.1098/rspb.2012.2686.

Danforth, B. (2007). Primer bees. *Current Biology, 17,* R156–R161. doi: 10.1016/j.cub.2007.01.025.

Dyer, A. G., Boyd-Gerny, S., McLoughlin, S., Rosa, M. G. P., Simonov, V., & Wong, B. B. M. (2012). Parallel evolution of angiosperm colour signals: Common evolutionary pressures linked to hymenopteran vision. *Proceedings of the Royal Society B Biological Sciences, 279,* 3606–3615. doi: 10.1098/rspb.2012.0827.

Friedman, W. E. (2009). The meaning of Darwin's "abominable mystery." *American Journal of Botany, 96,* 5–21. doi: 10.3732/ajb.0800150.

Hurd, P. D., Linsley, E. G., & Whitaker, T. W. (1971). Squash and gourd bees (*Peponapis, Xenoglossa*) and the origin of the cultivated *Cucurbita. Evolution, 25*(1), 218–234. doi: 10.2307/2406514.

Stebbins, G. L. (1970). Adaptive radiation of reproductive characteristics in angiosperms, I: Pollination mechanisms. *Annual Review of Ecology and Systematics, 1,* 307–326. https://doi.org/10.1146/annurev.es.01.110170.001515.

Floral Adaptations

Proctor, M., Yeo, P., & Lack, A. (1996). *The natural history of pollination* (Chapters 2, 5, 6, 14). Portland, Oregon: Timber Press.

Anatomical Adaptations

Dade, H. A. (1977). *Anatomy and dissection of the honeybee.* Cardiff, United Kingdom: International Bee Research Association.

Graham, J. (Ed.). (1992). *The hive and the honey bee* (Chapters 3, 8). Hamilton, Illinois: Dadant and Sons.

Harris, W. H. (1884). Religious Tract Society (https://www.biodiversitylibrary.org/page/21494767#page/119/mode/1up). In the public domain.

Snodgrass, R. E. (1956). *Anatomy of the honey bee.* Ithaca, New York: Comstock Publishing Associates.

Southwick, E. (1992). Physiology and social physiology of the honey bee. In J. Graham (Ed.), *The Hive and the Honey Bee* (pp. 171–196). Hamilton, Illinois: Dadant and Sons.

Winston, M. L. (1987). *The biology of the honey bee* (Chapter 3). Cambridge, Massachusetts: Harvard University Press.

Physiological Adaptations

von Frisch, K. (1971). *Bees: Their vision, chemical senses, and language* (Rev. ed.). Ithaca, New York: Cornell University Press.

Winston, M. L. (1987). *The biology of the honey bee* (Chapter 4). Cambridge, Massachusetts: Harvard University Press.

Learning

Page, R. E., Jr. (2013). *The spirit of the hive: The mechanisms of social evolution* (Chapter 5). Cambridge, Massachusetts: Harvard University Press.

Page, R. E., Scheiner, R., Erber, J., & Amdam, G. V. (2006). The development and evolution of division of labor and foraging specialization in a social insect (*Apis mellifera* L.). *Current Topics in Developmental Biology, 74*, 253–286. doi: 10.1016/S0070-2153(06)74008-X.

Scheiner, R., Barnert, M., & Erber, J. (2003). Variation in water and sucrose responsiveness during the foraging season affects proboscis extension learning in honey bees. *Apidologie, 34*(1), 67–72. https://doi.org/10.1051/apido:2002050.

Navigation

Cheeseman, J. F., Millar, C. D., Greggers, U., Lehmann, K., Pawley, M. D. M., Gallistel, C. R., Warman, G. R., & Menzel, R. (2014). Way-finding in displaced clock-shifted bees prove bees use a cognitive map. *Proceedings of the National Academy of Sciences USA, 111*(24), 8949–8954. https://doi.org/10.1073/pnas.1408039111.

Cheung, A., Collett, M., Collett, T. S., Dewar, A., Dyer, F., Graham, P., Mangan, M., Narendra, A., Philippides, A., Stürzl, W., Webb, B., Wystrach, A., & Zeil, J. (2014). Still no convincing evidence for cognitive map use by honey bees. *Proceedings of the National Academy of Sciences USA, 111*(42), E4396–E4397. https://doi.org/10.1073/pnas.1413581111.

Cruse, H., & Wehner, R. (2011). No need for a cognitive map: Decentralized memory for insect navigation. *PLoS Computational Biology, 7*, e100209. https://doi.org/10.1371/journal.pcbi.1002009.

Dance Language

Gould, J. L. (1976). The dance language controversy. *The Quarterly Review of Biology*, *51*, 211–244.

von Frisch, K. (1967). *The dance language and orientation of bees* (Chapters 5, 12, 13). Cambridge, Massachusetts: Belknap Press of Harvard University Press.

Honey Bees in the New World

Graham, J. (Ed.). (1992). *The hive and the honey bee* (Chapter 1). Hamilton, Illinois: Dadant and Sons.

Pollination of Agricultural Crops

McGregor, S. E. (1976). *Insect pollination of cultivated crop plants* (Chapter 1) (Agricultural Handbook No. 496, Agricultural Research Service). Washington, DC: US Department of Agriculture.

Almonds

Ferris, J. (2013, September 1). The mind-boggling math of migratory beekeeping. *Scientific American*. http://scientificamerican.com/article/migratory-beekeeping-mind-boggling-math/.

US Department of Agriculture, National Agricultural Statistics Service. (2018, May 10). *2018 California almond forecast*. https://www.nass.usda.gov/Statistics_by_State/California/Publications/Specialty_and_Other_Releases/Almond/Forecast/201805almpd.pdf.

Alfalfa

Piper, C. V., Evans, M. W., McKee, R., & Morse, W. J. (1914). *Alfalfa seed production, pollination studies* (US Department of Agriculture Bulletin No. 75). Washington, DC: US Government Printing Office.

Todd, F. E., & McGregor, S. E. (1960). The use of honey bees in the production of crops. *Annual Review of Entomology*, *5*(1), 265–278. doi: 10.1146/annurev.en.05.010160.001405.

The Decline of Honey Bees

Biesmeijer, J. C., Roberts, S. P. M., Reemer, M., Ohlemüller, R., Edwards, M., Peeters, T., Schaffers, A. P., Potts, S. G., Kleukers, R., Thomas, C. D., Settele, J., & Kunin, W. E. (2006). Parallel declines in pollinators and insect-pollinated plants in Britain and the Netherlands. *Science*, *313*, 351–354. doi: 10.1126/science.1127863.

Kraus, B., & Page, R. E. (1995). Effect of *Varroa jacobsoni* (Mesostigmata: Varroadae) on feral *Apis mellifera* (Hymenoptera: Apidae) in California. *Environmental Entomology*, *24*(6), 1473–1480.

Kraus, B., & Page, R. E. (1995). Population growth of *Varroa jacobsoni* Oud in Mediterranean climates of California. *Apidologie*, *26*(2), 149–157. https://doi.org/10.1051/apido:19950208.

Winston, M. L. (2014). *Bee time* (Chapter 4). Cambridge, Massachusetts: Harvard University Press.

How Bees "Paint" Their Local Environments

Seeley, T. D. (1995). *The wisdom of the hive: The social physiology of honey bee colonies* (Chapter 3). Cambridge, Massachusetts: Harvard University Press.

Colonies as Optimum Foragers

Seeley, T. D. (1995). *The wisdom of the hive: The social physiology of honey bee colonies* (Chapter 5). Cambridge, Massachusetts: Harvard University Press.

Individual Bees as Optimum Foragers

Harano, K., Mitsuhata-Asai, & Sasaki, M. (2014). Honey loading for pollen collection: regulation of crop content in honeybee pollen foragers on leaving hive. *Naturwissenschaften* 101(7):595-598. htpps://doi: 10.1007/s00114-014-1185-z

Page, R. E., Jr. (2013). *The spirit of the hive: The mechanisms of social evolution* (Chapter 9). Cambridge, Massachusetts: Harvard University Press.

Pankiw, T., Nelson, M., Page, R. E., & Fondrk, M. K. (2004). The communal crop: modulation of sucrose response thresholds of pre-foraging honey bees with incoming nectar quality. *Behavioral Ecology and Sociobiology*, *55*, 286–292.

Schmid-Hempel, P., Kacelnik, A., & Houston, A. I. (1985). Honeybees maximize efficiency by not filling the crop. *Behavioral Ecology and Sociobiology*, *17*(1), 61–66. https://doi.org/10.1007/BF00299430.

Regulation of Pollen Foraging

Page, R. E., Jr. (2013). *The spirit of the hive: The mechanisms of social evolution* (Chapter 9). Cambridge, Massachusetts: Harvard University Press.

Waddington, K. D., Nelson, C. M., & Page, R. E. (1998). Effects of pollen quality and genotype on the dance of foraging honey bees. *Animal Behavior*, *56*, 35–39.

Chapter 3

Dr. Seuss. (1955). *On beyond zebra.* New York: Random House, Inc.

Engineering the Nest

Darwin, C. D. (1998). *The origin of species.* New York, New York: The Modern Library (Original work published 1859).

Graham, J. (Ed.). (1992). *The hive and the honey bee* (Chapter 3). Hamilton, Illinois: Dadant and Sons.

Grinnel, J. (1917). The niche-relationships of the California thrasher. *Auk*, *34*, 427–433.

Hutchinson, G. (1959). Homage to Santa Rosalia or why are there so many kinds of animals? *The American Naturalist*, *93*, 145.

Krakauer, D. C., Page, K. M., & Erwin, D. H. (2009). Diversity, dilemmas, and monopolies of niche construction. *The American Naturalist*, *173*(1), 26–40. doi: 10.1086/593707.

Page, R. E. (1981). Protandrous reproduction in honey bees. *Environmental Entomology*, *10*(3), 359–362. https://doi.org/10.1093/ee/10.3.359.

Page, R. E., & Metcalf, R. A. (1984). A population investment sex ratio for the honey bee (*Apis mellifera* L.). *The American Naturalist*, *124*(5), 680–702. https://doi.org/10.1086/284306.

Seeley, T. D. (2010). *Honeybee democracy* (Chapter 3). Princeton, New Jersey: Princeton University Press.

Seeley, T. D., & Morse, R. A. (1976). The nest of the honey bee (*Apis mellifera* L.). *Insectes Sociaux*, *23*, 495–512. https://doi.org/10.1007/BF02223477.

Wilson, E.O. (1971). *The insect societies*. Cambridge, Massachusets: Belknap Press of Harvard University Press.

Winston, M. L. (1987). *The biology of the honey bee* (Chapter 5). Cambridge, Massachusetts: Harvard University Press.

Thermal Regulation

Winston, M. L. (1987). *The biology of the honey bee* (Chapter 7). Cambridge, Massachusetts: Harvard University Press.

Nest Hygiene

Kraus, B., & Page, R. E. (1995). Effect of *Varroa jacobsoni* (Mesostigmata: Varroadae) on feral *Apis mellifera* (Hymenoptera: Apidae) in California. *Environmental Entomology, 24*, 1473–1480.

Rothenbuhler, W. C. (1958). Genetics and breeding of the honey bee. *Annual Review of Entomology, 3*, 161–180. https://doi.org/10.1146/annurev.en.03.010158.001113.

Rothenbuhler, W. C. (1964). Behavior genetics of nest cleaning in honey bees. IV. Responses of F1 and backcross generations to disease-killed brood. *American Zoologist, 4*, 111–123.

Individual Immunity

Evans, J. D., Aronstein, K., Chen, Y. P., Hetru, C., Imler, J. L., Kanost, M., Thompson, G. J., Zou, Z., & Hultmark, D. (2006). Immune pathways and defence mechanisms in honey bees *Apis mellifera*. *Insect Molecular Biology, 15*, 645–656. doi: 10.1111/j.1365-2583.2006.00682.x.

Tsakas, S., & Marmaras, V. J. (2010). Insect immunity and its signaling: An overview. *Invertebrate Survival Journal, 7*(2), 228–238.

Social Immunity

Papachristoforou, A. R., Zaferiridou, G., Theophilidis, G., Gartnery, L., & Arnold, G. (2007). Smothered to death: Hornets asphyxiated by honeybees. *Current Biology, 17*(18), R795–R796. https://doi.org/10.1016/j.cub.2007.07.033.

Van Meyel, S., Körner, M., & Meunier, J. (2018). Social immunity: Why we should study its nature, evolution and function across all social systems. *Current Opinion in Insect Science, 28*, 1–7. https://doi.org/10.1016/j.cois.2018.03.004.

Selection for Disease Resistance

Kraus, B., & Page, R. E. (1998). Parasites, pathogens, and polyandry in social insects. *The American Naturalist, 151*(4), 383–391. doi: 10.1086/286126.

Page, R. E., & Guzmán-Novoa, E. (1997). The genetic basis of disease resistance. In R. A. Morse (Ed.), *Honey bee pests, predators, and diseases* (pp. 471–491). Medina, Ohio: A. I. Root Co.

Sutter, G. R., Rothenbuhler, W. C., & Raun, E. S. (1968). Resistance to American foulbrood in honey bees VII. Growth of susceptible and resistant larvae. *Journal of Invertebrate Pathology, 12*(1), 25–28. https://doi.org/10.1016/0022-2011(68)90239-5.

Social Recognition

Breed, M. D. (2014). Kin and nestmate recognition: The influence of W. D. Hamilton on 50 years of research. *Animal Behaviour, 92*, 271–279. https://doi.org/10.1016/j.anbehav.2014.02.030.

Breed, M. D., Cook, C. N., McCreery, H. F., & Rodriguez, M. (2015). Nestmate recognition in eusocial insects: The honeybee as a model system. In L. Aquiloni & E. Tricarico (Eds.), *Social recognition in invertebrates* (pp. 147–164). Springer International Publishing.

Nest Defense

Breed, M. D., Guzmán-Novoa, E., & Hunt, G. J. (2004). Defensive behavior of honey bees: Organization, genetics, and comparisons with other bees. *Annual Review of Entomology, 49*, 271–298. doi: 10.1146/annurev.ento.49.061802.123155.

Dade, H. A. (1977). *Anatomy and dissection of the honeybee.* Cardiff, United Kingdom: International Bee Research Association.

Graham, J. (Ed.). (1992). *The hive and the honey bee* (Chapters 1 and 2). Dadant and Sons.

Guzmán-Novoa, E., & Page, R. E. (1994). The impact of Africanized bees on Mexican beekeeping. *American Bee Journal, 134*(2), 101–106.

Impact of Africanized Honey Bees

Baum, K. A., Rubink, W. L., Pinto, M. A., & Coulson, R. N. (2005). Spatial and temporal distribution and nest site characteristics of feral honey bee (Hymenoptera: Apidae) colonies in a coastal prairie landscape. *Environmental Entomology, 34*(3), 610–618. https://doi.org/10.1603/0046-225X-34.3.610.

Guzmán-Novoa, E., & Page, R. E. (1994). Genetic dominance and worker interactions affect honeybee colony defense. *Behavioral Ecology, 5*(1), 91–97. https://doi.org/10.1093/beheco/5.1.91.

Guzmán-Novoa, E., Page, R. E., & Fondrk, M. K. (1994). Morphometric techniques do not detect intermediate and low levels of Africanization in honey bee (Hymenoptera: Apidae) colonies. *Annals of the Entomological Society of America, 87*(5), 507–515. https://doi.org/10.1093/aesa/87.5.507.

Michener, C. D. (1975). The Brazilian bee problem. *Annual Review of Entomology, 20*, 399–416. https://doi.org/10.1146/annurev.en.20.010175.002151.

Nogueira-Neto, P. (1964). The spread of a fierce African bee in Brazil. *Bee World, 45*, 119–121. https://doi.org/10.1080/0005772X.1964.11097060.

Roubik, D. W., & Villanueva-Gutiérrez, R. (2009). Invasive Africanized honey bee impact on native solitary bees: A pollen resource and trap nest analysis. *Biological Journal of the Linnean Society, 98*, 152–160. https://doi.org/10.1111/j.1095-8312.2009.01275.x.

Taylor, O. R. (1985). African bees: Potential impact in the United States. *Bulletin of the Entomological Society of America, 31*, 15–24. https://doi.org/10.1093/besa/31.4.15.

Niche Construction or Environmental Engineering?

Gary, N. E. (1992). Activities and behavior of honey bees. In J. Graham (Ed.), *The hive and the honey bee* (pp. 185–264). Hamilton, Illinois: Dadant and Sons.

Laland, K. N., Odling-Smee, F. J., & Feldman, M. W. (1999). Evolutionary consequences of niche construction and their implications for ecology. *Proceedings of the National Academy of Sciences USA, 96*(18), 10242–10247. https://doi.org/10.1073/pnas.96.18.10242.

Odling-Smee, F. J., Laland, K. N., & Feldman, M. W. (2003). *Niche construction: The neglected process in evolution* (Chapter 1). Princeton, New Jersey: Princeton University Press.

Conditions That Enhance the Evolution of Niche Construction Traits

Corby-Harris, V., Snyder, L., Meador, C., & Ayotte, T. (2018). Honey bee (*Apis mellifera*) nurses do not consume pollens based on their nutritional quality. *PLoS ONE, 13*(1), e0191050. doi: 10.1371/journal.pone.0191050.

Gould, J. L. (1976). The dance-language controversy. *The Quarterly Review of Biology, 51*(2), 211–244. https://doi.org/10.1086/409309.

Graham, J. (Ed.). (1992). *The hive and the honey bee* (Chapter 8). Hamilton, Illinois: Dadant and Sons.

Krakauer, D. C., Page, K. M., & Erwin, D. H. (2009). Diversity, dilemmas, and monopolies of niche construction. *The American Naturalist, 173*(1), 26–40. doi: 10.1086/593707.

Page, R. E., Erber, J., & Fondrk, M. K. (1998). The effect of genotype on response thresholds to sucrose and foraging behavior of honey bees (*Apis mellifera* L.). *Journal of Comparative Physiology A, 182*, 489–500. https://doi.org/10.1007/s003590050196.

Chapter 4

The Social Contract

Constitution Society. (2007). The social contract and constitutional republics. http://www.constitution.org/soclcont.htm.

Friend, C. (n.d.). Social contract theories. In *Internet Encyclopedia of Philosophy*. http://www.iep.utm.edu/soc-cont/.

Meier, H. (2017). *Political philosophy and the challenge of revealed religion*. (R. Berman, Trans.; Chapter 3). Chicago, Illinois: University of Chicago Press.

Rousseau, J.-J. (2016). The social contract In *The Jean-Jacques Rousseau collection*. First Rate Publishers (G. D. H. Cole, J. Steeg, & S. W. Orson, Trans.) (Original work published 1762).

Wilson, E. O. (2012). *The social conquest of earth*. New York, New York: Liveright Publishing.

Social States of Bees

Batra, S.W. T. (1968). Behavior of some social and solitary halictine bees within their nests: a comparative study (Hymenoptera: Halictidae). *Journal of the Kansas Entomological Society, 41*, 120–133.

Michener, C. D. (1974). *The social behavior of the bees*. Cambridge, Massachusetts: Belknap Press of Harvard University Press.

Social Evolution

Hamilton, W. D. (1964a). The genetical evolution of social behavior I. *Journal of Theoretical Biology, 7*, 1–16. https://doi.org/10.1016/0022-5193(64)90038-4.

Hamilton, W. D. (1964b). The genetical evolution of social behavior II. *Journal of Theoretical Biology, 7*, 17–52. https://doi.org/10.1016/0022-5193(64)90039-6.

Hamilton, W. D. (1972). Altruism and related phenomena, mainly in social insects. *Annual Review of Ecology and Systematics, 3*, 193–232. https://doi.org/10.1146/annurev.es.03.110172.001205.

Trivers, R. L., & Hare, H. (1972). Haplodiploidy and the evolution of the social insects. *Science, 191*, 249–263. doi: 10.1126/science.1108197.

Sex Ratios

Bourke, A. F. (2015). Sex investment ratios in eusocial Hymenoptera support inclusive fitness theory. *Journal of Evolutionary Biology, 28*, 2106–2111. doi: 10.1111/jeb.12710.

Del Fabro, A., Driul, L., Anis, O., Londero, A. P., Bertozzi, S., Bortollo, L., & Marchesoni, D. (2011). Fetal gender ratio in recurrent miscarriages. *International Journal of Women's Health, 3*, 213–217. doi: 10.2147/IJWH.S20557.

Fisher, R.A. (1930). *The Genetical Theory of Natural Selection*. Oxford: Clarendon Press.

Gardner, A., Alpedrinha, J., & West, S. A. (2012). Haplodiploidy and the evolution of eusociality: Split sex ratios. *The American Naturalist, 179*(4), 240–256. doi: 10.1086/663683.

Meunier, J., West, S. A., & Chapuisat, M. (2008). Split sex ratios in the social Hymenoptera: A meta-analysis. *Behavioral Ecology, 19*(2), 382–390. doi: https://doi.org/10.1093/beheco/arm143.

Queller, D. C. (2006). Sex ratios and social evolution. *Current Biology, 16*(17), 664–668. https://doi.org/10.1016/j.cub.2006.08.017.

Schwarz, M. P. (1994). Female-biased sex ratios in a facultatively social bee and their implications for social evolution. *Evolution, 48*(5), 1684–1697. doi: 10.1111/j.1558-5646.1994.tb02205.x.

Wikipedia. (n.d.). *Human sex ratio*. https://en.wikipedia.org/wiki/Human_sex_ratio

Chapter 5

The Wheeler Superorganism

Butler, C. (2017). *The Feminine Monarchie*, edited by John Owen. Yorkshire, UK: Northern Bee Books.

Darwin, C. D. (1998). *The origin of species*. New York, New York: The Modern Library (Original work published 1859).

Hobbes, T. (2002), *Leviathan*. Project Gutenberg. Retrieved August 16, 2019 (http://www.gutenberg.org/ebooks/3207). In the public domain.

Maeterlinck, M. (1913) *The life of the bee*. New York, New York: Dodd, Mead, & Company.

Spencer, H. (1860). The social organism. In: Essays: Scientific, Political, and Speculative. London, England. (Essay first published in The Westminster Review, January 1860).

Spencer, H. (1864). The principles of biology (vol 1). London, England: Williams and Norgate.

Marais, E. (2009). *The soul of the white ant*. New York, New York: New York University Press.

Wheeler, W. M. (1911). The ant colony as an organism. *Journal of Morphology, 22*(2), 307–325. https://doi.org/10.1002/jmor.1050220206.

Wheeler, W. M. (1923). *Social life among the insects*. New York, New York: Harcourt Brace and Company.

Wheeler, W. M. (1928). *The social insects*. New York, New York: Harcourt Brace and Company.

The Decline and Resurrection of the Superorganism

Boomsma, J. J., & Gawne, R. (2017). Superorganismality and caste differentiation as points of no return: How the major evolutionary transitions were lost in translation.

Biological Reviews of the Cambridge Philosophical Society, 93, 28–54. https://doi.org/10.1111/brv.12330.

Gardner, A., & Grafen, A. (2009). Capturing the superorganism: A formal theory of group adaptation. *Journal of Evolutionary Biology*, 22, 659–671. doi: 10.1111/j.1420-9101.2008.01681.x.

Hölldobler, B., & Wilson, E. O. (2008). *The superorganism: The beauty, elegance, and strangeness of insect societies.* New York, New York: W.W. Norton and Company.

Wilson, D. S., & Sober, E. (1989). Reviving the superorganism. *Journal of Theoretical Biology*, 136(3), 337–356. doi: 10.1016/s0022-5193(89)80169-9.

Chapter 6

Reproductive Competition

Koeniger, N. (1970). Factors determining the laying of drone and worker eggs by the queen honeybee. *Bee World*, 51(4), 166–169. https://doi.org/10.1080/0005772X.1970.11097324.

Naeger, N. L., Peso, M., Even, N., Barron, A. B., & Robinson, G. E. (2013). Altruistic behavior by egg-laying worker honeybees. *Current Biology*, 23(16), 1574–1578. doi: 10.1016/j.cub.2013.06.045.

Page, R. E. (1981). Protandrous reproduction in honey bees. *Environmental Entomology*, 10(3), 359–362. https://doi.org/10.1093/ee/10.3.359.

Page, R. E. (1986). Sperm utilization in social insects. *Annual Review of Entomology*, 31, 297–320. https://doi.org/10.1146/annurev.en.31.010186.001501.

Page, R. E., & Erickson, E. H. (1986). Kin recognition during emergency queen rearing by worker honey bees (Hymenoptera: Apidae). *Annals of the Entomological Society of America*, 79(3), 460–467. https://doi.org/10.1093/aesa/79.3.460.

Page, R. E., & Erickson, E. H. (1988). Reproduction by worker honey bees (*Apis mellifera* L.). *Behavioral Ecology and Sociobiology*, 23(2), 117–126. https://doi.org/10.1007/BF00299895.

Page, R. E., & Robinson, G. E. (1994). Reproductive competition in queenless honey bee colonies (*Apis mellifera* L.). *Behavioral Ecology and Sociobiology*, 35(2), 99–107. https://doi.org/10.1007/BF00171499.

Robinson, G. E., Page, R. E., & Fondrk, M. K. (1990). Intracolonial behavioral variation in worker oviposition, oophagy, and larval care in queenless honey bee colonies. *Behavioral Ecology and Sociobiology*, 26(5), 315–323. https://doi.org/10.1007/BF00171096.

The Cape Honey Bee

Hepburn, H. R. (2001). The enigmatic Cape honey bee, *Apis mellifera capensis*. *Bee World*, 82, 181–191. https://doi.org/10.1080/0005772X.2001.11099525.

Hepburn, H. R., & Crewe, R. M. (1991). Portrait of the Cape honeybee, *Apis mellifera capensis*. *Apidologie*, 22(6), 567–580. https://doi.org/10.1051/apido:19910601.

Oldroyd, B. P., Allsopp, M. H., Gloag, R. S., Lim, J., Jordan, L. A., & Beekman, M. (2008). Thelytokous parthenogenesis in unmated queen honeybees (*Apis mellifera capensis*): Central fusion and high recombination rates. *Genetics*, 180(1), 359–366. doi: 10.1534/genetics.108.090415.

Ruttner, F. (1977). The problem of the Cape bee (*Apis mellifera capensis* Eschotz): Parthenogensis–size of population–evolution. *Apidologie*, *8*(3), 281–294. https://doi.org/10.1051/apido:19770305.

Tucker, K. W. (1957). Automictic parthenogenesis in the honey bee. *Genetics*, *43*(3), 299–316.

Wallberg, A., Pirk, C. W., Allsopp, M. H., & Webster, M. T. (2016). Identification of multiple loci associated with social parasitism in honeybees. *PLoS Genetics*. doi: 10.1371/journal.pgen.1006097.

Disease in the Superorganism

Amdam, G. V., & Seehus, S.-C. (2006). Order, disorder, death: Lessons from a superorganism. *Advances in Cancer Research*, *95*, 31–60. doi: 10.1016/S0065-230X(06)95002-7.

The Honey Bee as a Wheeler Superorganism

Moritz, R. F. A., & Southwick, E. E. (1992). *Bees as superorganisms: An evolutionary reality*. Berlin, Germany: Springer Verlag.

Page, R. E., & Mitchell, S. D. (1993). The superorganism: New perspective or tired metaphor? *Trends in Ecology and Evolution*, *8*(7), 265–266.

Seeley, T. D. (1989). The honey bee colony as a superorganism. *American Scientist*, *77*(6), 546–553.

Chapters 7 and 8

Hölldobler, B., & Wilson, E. O. (2008). *The superorganism: The beauty, elegance, and strangeness of insect societies*. New York, New York: W.W. Norton and Company.

Page, R. E. (2013). *The spirit of the hive: The mechanisms of social evolution*. Cambridge, Massachusetts: Harvard University Press.

How to Make a Worker

Dedej, S., Hartfelder, K., Aumeier, P., Rosenkranz, P., & Engels, W. (1998). Caste determination is a sequential process: Effect of larval age at grafting on ovariole number, hind leg size and cephalic volatiles in the honey bee (*Apis mellifera carnica*). *Journal of Apicultural Research*, *37*(3), 183–190. https://doi.org/10.1080/00218839.1998.11100970.

Jianke, L., & Aiping, W. (2005). Comprehensive technology for maximizing royal jelly production. *American Bee Journal*, *145*, 661–664.

Worker Ovaries and Behavior

Ihle, K. E., Page, R. E., Frederick, K., Fondrk, M. K., & Amdam, G. V. (2010). Genotypic effect on regulation of behaviour by *vitellogenin* supports reproductive origin of honeybee foraging bias. *Animal Behaviour*, *79*(1), 1001–1006. doi: 10.1016/j.anbehav.2010.02.009.

Nelson, C. M., Ihle, K. E., Fondrk, M. K., Page, R. E., & Amdam, G. V. (2007). The gene *vitellogenin* has multiple coordinating effects on social organization. *PLoS Biology*, *5*, 673–677. https://doi.org/10.1371/journal.pbio.0050062.

Robinson, G. E., Page, R. E., & Fondrk, M. K. (1990). Intracolonial behavioral variation in worker oviposition, oophagy, and larval care in queenless honey bee colonies. *Behavioral Ecology and Sociobiology*, *26*(5), 315–323. https://doi.org/10.1007/BF00171096.

Tsuruda, J. M., Amdam, G. V., & Page, R. E. (2008). Sensory response system of social behavior tied to female reproductive traits. *PLoS ONE, 3*, e3397. https://doi.org/10.1371/journal.pone.0003397.

Wang, Y., Kaftanoglu, O., Fondrk, M. K., & Page, R. E. (2014). Nurse bee behaviour manipulates worker honeybee (*Apis mellifera* L.) reproductive development. *Animal Behaviour, 92*, 253–261. doi: https://doi.org/10.1016/j.anbehav.2014.02.012.

Evolution of Worker and Queen Development

Amdam, G. V., Norberg, K., Hagen, A., & Omholt, S. W. (2003). Social exploitation of vitellogenin. *Proceedings of the National Academy of Sciences USA, 100*(4), 1799e1802. https://doi.org/10.1073/pnas.0333979100.

Amdam, G. V., & Page, R. E. (2010). The developmental genetics and physiology of honeybee societies. *Animal Behaviour, 79*(5), 973–980. doi: 10.1016/j.anbehav.2010.02.007.

Leimar, O., Hartfelder, K., Laubichler, M. D., & Page, R. E. (2012). Development and evolution of caste dimorphism in honey bees—A modeling approach. *Ecology and Evolution, 2*(12), 3098–3109. doi: 10.1002/ece3.414.

How Selection on Colonies Builds Superorganism Traits

Amdam, G. V., Norberg, K., Fondrk, M. K., & Page, R. E. (2004). Reproductive ground plan may mediate colony-level selection effects on individual foraging behavior in honey bees. *Proceedings of the National Academy of Sciences USA, 101*(31), 11350–11355. https://doi.org/10.1073/pnas.0403073101.

Gardner, A., & Grafen, A. (2009). Capturing the superorganism: A formal theory of group adaptation. *Journal of Evolutionary Biology, 22*, 659–671. doi: 10.1111/j.1420-9101.2008.01681.x.

Page, R. E., & Amdam, G. V. (2007). The making of a social insect: Developmental architectures of social design. *Bioessays, 29*(4), 334–343. doi: 10.1002/bies.20549.

Page, R. E., & Erber, J. (2002). Levels of behavioral organization and the evolution of division of labor. *Naturwissenschaften, 89*(3), 91–106. https://doi.org/10.1007/s00114-002-0299-x.

Page, R. E., Scheiner, R., Erber, J., & Amdam, G. V. (2006). The development and evolution of division of labor and foraging specialization in a social insect (*Apis mellifera* L.). *Current Topics in Developmental Biology, 74*, 253–286. doi: 10.1016/S0070-2153(06)74008-X.

Temporal Changes in Behavior

Amdam, G. V., & Omholt, S. W. (2003). The hive bee to forager transition in honeybee colonies: The double repressor hypothesis. *Journal of Theoretical Biology, 223*(4), 451–464. doi: 10.1016/s0022-5193(03)00121-8.

Calderone, N. W., & Page, R. E. (1996). Temporal polyethism and behavioural canalization in the honey bee, *Apis mellifera*. *Animal Behaviour, 51*, 631–643. https://doi.org/10.1006/anbe.1996.0068.

Page, R. E. (1997). The evolution of insect societies. *Endeavour, 21*, 114–120.

Response Thresholds and Social Organization

Beekman, M. K., Preece, K., & Schaerf, T. M. (2016). Dancing for their supper: Do honeybees adjust their recruitment dance in response to the protein content of pollen? *Insectes Sociaux, 63*(1), 117–126. doi: 10.1007/s00040-015-0443-1.

Muth, F., Papaj, D. R., & Leonard, A. S. (2016). Bees remember flowers for more than one reason: Pollen mediates associative learning. *Animal Behaviour, 111*, 93–100. https://doi.org/10.1016/j.anbehav.2015.09.029.

A Selection Experiment

Dobzhansky, T. (1973). Nothing in biology makes sense except in the light of evolution. *American Biology Teacher, 35*(3), 125–129. doi: 10.2307/4444260.

Estoup, A., Garnery, L., Solignac, M., & Cornuet, J. M. (1995). Microsatellite variation in honey bee (*Apis mellifera* L.) populations: Hierarchical genetic structure and test of the infinite allele and stepwise mutation models. *Genetics, 140*(2), 679–695.

Page, R. E., Erickson, E. H., & Laidlaw, H. H. (1982). A closed population breeding program for honey bees. *American Bee Journal, 122*, 350, 351, 354, 355.

Page, R. E., & Fondrk, M. K. (1995). The effects of colony-level selection on the social organization of honey bee (*Apis mellifera* L.) colonies: Colony-level components of pollen hoarding. *Behavioral Ecology and Sociobiology, 36*, 135–144. https://doi.org/10.1007/BF00170718.

Severson, D. L., Page, R. E., & Erickson, E. H., Jr. (1986). Closed population breeding in honey bees: A report on its practical application. *American Bee Journal, 126*, 93–94.

Larval Development—A Sociogenome

Linksvayer, R. A., Fondrk, M. K., & Page, R. E. (2009). Honeybee social regulatory networks are shaped by colony-level selection. *The American Naturalist, 173*(3), E99–E107. doi: 10.1086/596527.

Linksvayer, T. A., Kaftanoglu, O., Akyol, E., Blatch, S., Amdam, G. V., & Page, R. E. (2011). Larval and nurse worker control of developmental plasticity and the evolution of honey bee queen–worker dimorphism. *Journal of Evolutionary Biology, 24*(9), 1939–1948. doi: 10.1111/j.1420-9101.2011.02331.x.

Regulation of Stored Pollen

Fewell, J. H., & Winston, M. L. (1992). Colony state and regulation of pollen foraging in the honey bee, *Apis mellifera* L. *Behavioral Ecology and Sociobiology, 30*(6), 387–393. https://doi.org/10.1007/BF00176173.

Grozinger, C. M., Fischer, P., & Hampton, J. E. (2007). Uncoupling primer and releaser responses to pheromone in honey bees. *Naturwissenschaften, 94*, 375–379. doi: 10.1007/s00114-006-0197-8.

Rueppell, O., Bachelier, C., Fondrk, M. K., & Page, R. E. (2007). Regulation of life history regulates lifespan of worker honey bees. *Experimental Gerontology, 42*(10), 1020–1032. doi: 10.1016/j.exger.2007.06.002.

Traynor, K. S., Le Conte, Y., & Page, R. E. (2014). Queen and young larval pheromones impact nursing and reproductive physiology of honey bee (*Apis mellifera*) workers. *Behavioral Ecology and Sociobiology, 68*(12), 2059–2073. doi: 10.1007/s00265-014-1811-y.

Traynor, K. S., Le Conte, Y., & Page, R. E. (2015). Age matters: Pheromone profiles of larvae differentially influence foraging behaviour in the honeybee, *Apis mellifera*. *Animal Behaviour, 99*, 1–8. https://doi.org/10.1016/j.anbehav.2014.10.009.

Traynor, K. S., Wang, Y., Brent, C. S., Amdam, G. V., & Page, R. E. (2017). Young and old (*Apis mellifera*) larvae differentially prime developmental maturation of their caregivers. *Animal Behaviour, 124*, 193–202. https://doi.org/10.1016/j.anbehav.2016.12.019.

Genes

Page, R. E., Rüppell, O., & Amdam, G. V. (2012). Genetics of reproduction and regulation of honey bee (*Apis mellifera* L.) social behavior. *Annual Review of Genetics, 46*, 97–119. doi: 10.1146/annurev-genet-110711-155610.

Rueppell, O., Amdam, G. V., Page, R. E., & Carey, J. R. (2004). From genes to societies. *Science of Aging Knowledge Environment, 5*, 1–11. doi: 10.1126/sageke.2004.5.pe5.

Rueppell, O., Chandra, S., Pankiw, T., Fondrk, M. K., Beye, M., Hunt, G., & Page, R. E. (2006). The genetic architecture of sucrose responsiveness in the honey bee (*Apis mellifera* L.). *Genetics, 172*(1), 243–251. doi: 10.1534/genetics.105.046490.

Chapter 9

The Song of the Queen

White, E. B. (1945, December 1). Song of the queen bee. *New Yorker*.

Coming Out

Chaline, N., Martin, S. J., & Ratnieks, F. L. W. (2005). Absence of nepotism toward imprisoned young queens during swarming in the honey bee. *Behavioral Ecology, 16*(2), 403–409. https://doi.org/10.1093/beheco/ari003.

Grooters, H. J. (1987). Influences of queen piping and worker behavior on the timing of emergence of honey bee queens. *Insectes Sociaux, 34*(3), 181–193. https://doi.org/10.1007/BF02224083.

Seeley, T. D. (2010). *Honeybee democracy* (Chapter 7). Princeton, New Jersey: Princeton University Press.

Simpson, J., & Cherry, S. M. (1969). Queen confinement, queen piping and swarming in *Apis mellifera* colonies. *Animal Behaviour, 17*(2), 271–278. https://doi.org/10.1016/0003-3472(69)90012-8.

Simpson, J., & Greenwood, S. P. (1974). Influence of artificial queen-piping sound on the tendency of honeybee, *Apis mellifera*, colonies to swarm. *Insectes Sociaux, 21*(3), 283–287. https://doi.org/10.1007/BF02226919.

Winston, M. L. (1987). *The biology of the honey bee* (Chapter 11). Cambridge, Massachusetts: Harvard University Press.

The Mating Flight

Gary, N. E. (1992). Activities and behavior of honey bees. In J. Graham (Ed.), *The hive and the honey bee* (pp. 185–264). Hamilton, Illinois: Dadant and Sons.

Koeniger, G. (1986). Mating sign and multiple mating in the honeybee. *Bee World, 67*, 141–150. doi: 10.1080/0005772X.1986.11098892.

Loper, G. M. (1992). What do we really know about drone flight behavior? *Bee World, 73*, 198–203. https://doi.org/10.1080/0005772X.1992.11099138.

Loper, G. M., Wolf, W. W., & Taylor, O. R. (1992). Honey bee drone flyways and congregation areas—Radar observations. *Journal of the Kansas Entomological Society, 65*(3), 223–230.

Lopez-Uribe, M. M., Appler, R. H., Youngsteadt, E., Dunn, R. R., Frank, S. D., & Tarpy, D. R. (2017). Higher immunocompetence is associated with higher genetic diversity in feral honey bee colonies (*Apis mellifera*). *Conservation Genetics, 18*, 659–666. doi: 10.1007/s10592-017-0942-x.

Ruttner, F. (1956). The mating of the honeybee. *Bee World, 37*, 3–15. https://doi.org/10.1080/0005772X.1956.11094913.

Sandoz, J.-C., Deisig, N., de Brito Sanchez, M. G., & Giurfa, M. (2007). Understanding the logics of pheromone processing in the honeybee brain: From labeled-lines to across-fiber patterns. *Frontiers in Behavioral Neuroscience, 1*, 5. doi: 10.3389/neuro.08.005.2007.

Taber, S. (1955). Sperm distribution in the spermathecal of multiple-mated queen honey bees. *Journal of Economic Entomology, 48*(5), 522–525. https://doi.org/10.1093/jee/48.5.522.

Taber, S., & Wendell, J. (1958). Concerning the number of times queen bees mate. *Journal of Economic Entomology, 51*(6), 786–789. https://doi.org/10.1093/jee/51.6.786.

Tarpy, D. R., & Page, R. E. (2000). No behavioral control over mating frequency in queen honey bees (*Apis mellifera* L.): Implications for the evolution of extreme polyandry. *The American Naturalist, 155*(6), 820–827. doi: 10.1086/303358.

Villar, G., Wolfson, M. D., Hefetz, A., & Grozinger, C. M. (2018). Evaluating the role of drone-produced chemical signals in mediating social interactions in honey bees (*Apis mellifera*). *Journal of Chemical Ecology, 44*(1), 1–8. doi: 10.1007/s10886-017-0912-2.

Woyke, J. (2011). The mating sign of queen bees originates from two drones and the process of multiple mating in honey bees. *Journal of Apicultural Research, 50*, 272–283. https://doi.org/10.3896/IBRA.1.50.4.04.

Woyke, J. (2016). Not the honey bee (*Apis mellifera* L.) queen, but the drone determines the termination of the nuptial flight and the onset of oviposition—The polemics, abnegations, corrections, and supplements. *Journal of Apicultural Science, 60*(2), 25–40. https://doi.org/10.1515/jas-2016-0032.

Sperm Acquisition and Use

Laidlaw, H. H. (1987). Instrumental insemination of honeybee queens: Its origin and development. *Bee World, 68*, 17–36. https://doi.org/10.1080/0005772X.1987.11098905.

Laidlaw, H. H., & Page, R. E. (1984). Polyandry in honey bees (*Apis mellifera* L.): Sperm utilization and intracolony genetic relationships. *Genetics, 108*(4), 985–997.

Page, R. E. (1986). Sperm utilization in social insects. *Annual Review of Entomology, 31*, 297–320. https://doi.org/10.1146/annurev.en.31.010186.001501.

Page, R. E., Kimsey, R. B., & Laidlaw, H. H. (1984). Migration and dispersal of spermatozoa in spermathecae of queen honeybees (*Apis mellifera* L.). *Experientia, 40*(2), 182–184. https://doi.org/10.1007/BF01963589.

Page, R. E., & Laidlaw, H. H. (1992). Honey bee genetics and breeding. In J. Graham (Ed.), *The hive and the honey bee* (pp. 235–267). Hamilton, Illinois: Dadant and Sons.

Page, R. E., & Metcalf, R. A. (1982). Multiple mating, sperm utilization, and social evolution. *The American Naturalist, 119*(2), 263–281. https://doi.org/10.1086/283907.

Genetic Resources for the Bee

Honeybee Genome Sequencing Consortium. (2006). Insights into social insects from the genome of the honeybee *Apis mellifera*. *Nature, 443*, 931–949. doi: 10.1038/nature05260.

Roth, A., Vleurinck, C., Netschitailo, O., Otte, M., Kaftanoglu, O., Page, R. E., & Beye, M. (2019). A genetic switch for worker nutrition-mediated traits in honeybees. *PLoS Biology, 17*(3), e3000171. https://doi.org/10.1371/journal.pbio.3000171.

Reply
Laidlaw, H. H. (1948, December 5). Response. *San Francisco Chronicle.*

Epilogue

Mitchell, S. D. (2003). *Biological complexity and integrative pluralism* (Chapter 4.2). Cambridge, United Kingdom: Cambridge University Press.
Thornhill, R. (1980). Pape [*sic*] in *Panorpa* scorpionflies and a general rape hypothesis. *Animal Behaviour, 28,* 52–59.

Index